Virtual Assistant

Edited by Ali Soofastaei

Published in London, United Kingdom

IntechOpen

Supporting open minds since 2005

Virtual Assistant
http://dx.doi.org/10.5772/intechopen.91579
Edited by Ali Soofastaei

Contributors
Abhishek Kaul, Moruf Akin Adebowale, Adriana Stan, Beáta Lőrincz, Musa Alhaji Ibrahim, Yusuf Şahin, Margherita Mori, Simon See, Aik Beng Ng, Zhangsheng Lai, Shaowei Lin, Ali Soofastaei, Auwal Ibrahim, Auwalu Yusuf Gidado, Mukhtar Nuhu Yahya

First published in London, United Kingdom, 2021 by IntechOpen
IntechOpen is the global imprint of INTECHOPEN LIMITED, registered in England and Wales, registration number: 11086078, 5 Princes Gate Court, London, SW7 2QJ, United Kingdom
Printed in Croatia

British Library Cataloguing-in-Publication Data
A catalogue record for this book is available from the British Library

Additional hard and PDF copies can be obtained from orders@intechopen.com

Virtual Assistant
Edited by Ali Soofastaei
p. cm.
Print ISBN 978-1-83968-807-2
Online ISBN 978-1-83968-808-9
eBook (PDF) ISBN 978-1-83968-809-6

For EU product safety concerns:
IN TECH d.o.o., Prolaz Marije Krucifikse Kozulić 3, 51000 Rijeka, Croatia, info@intechopen.com or visit our website at intechopen.com.

Meet the editor

Dr. Ali Soofastaei leads innovative industrial projects in artificial intelligence (AI) applications to improve safety, productivity, and energy efficiency and reduce maintenance costs. He holds a Bachelor of Engineering in Mechanical Engineering and has an in-depth understanding of energy management (EM) and equipment maintenance solutions (EMS). The extensive research he conducted on AI and value engineering (VE) methods while completing his Master of Engineering also provided him with expertise in applying advanced analytics in EM and EMS. Dr. Soofastaei completed his Ph.D. at The University of Queensland (UQ), Australia, in the field of AI applications in mining engineering, where he led a revolution in the use of deep learning (DL) and AI methods to increase energy efficiency, reduce operation and maintenance costs, and reduce greenhouse gas emissions in surface mines. In addition, as a Postdoctoral Research Fellow, Dr. Soofastaei has provided practical guidance to undergraduate and postgraduate students in mechanical and mining engineering and information technology. In the past fifteen years, Dr. Soofastaei has conducted various research studies in academic and industrial environments. As a result, he has acquired in-depth knowledge of energy efficiency opportunities (EEO), VE, and advanced analytics. He is an expert in using DL and AI methods in data analysis to develop predictive, optimization, and decision-making models of complex systems. Dr. Soofastaei has been involved in industrial research and development projects in several industries, including oil and gas (Royal Dutch Shell); steel (Danieli); and mining (BHP, Rio Tinto, Anglo American, and Vale). His extensive practical experience in the industry has equipped him to work with complex industrial problems in highly technical and multi-disciplinary teams. Dr. Soofastaei has more than ten years of academic experience as an assistant professor and leader of global research activities. His research and development projects have been published in international journals and keynote presentations. He has presented his practical achievements at conferences in the United States, Europe, Asia, and Australia. Dr. Soofastaei has founded Soofastaei-Publications, Soofastaei-Educations, and Soofastaei-Businesses institutes focusing on digital transformation and advanced analytics to provide not only the required materials and references for the 4.0 industrial digital revolution but also train the new generation of specialists and industrial managers who are interested in studying and working in the field of advanced applied analytics and AI.

Contents

Preface

Intelligent machines and computers have been developed to reduce time consumption and human efforts, improve safety, and increase the quality of products. Having virtual assistants (VAs) can play a critical role in using intelligent machines practically to complete projects efficiently. With the fast-growing technologies in the field, we have finally reached a stage where almost all the people living in developed and developing countries have access to these high technologies. However, this is just the starting point, and a long journey is ahead because future developments are taking a more advanced route in the shape of artificial intelligence (AI). Intelligent VAs use AI, and any improvement in AI potentially can develop VA-related technologies.

The main reason for using VAs is to replace humans with machines for completing repeatable tasks. This approach can give freedom to people to improve their innovations and find optimum solutions for personal and professional challenges. Previously, machines were doing what they were programmed to do, but now with AI, devices can think and behave like humans. This opportunity can help people to trust machines as VAs.

High-tech giants like Apple, Amazon, Google, Microsoft, Deloitte, Accenture, and IBM are very involved in research to develop the knowledge that has produced the new generation of VAs. Although VAs will form our future, we need to know how they affect our work and lifestyle. Thus, this book gives readers a glimpse of the role of VAs in shaping the future world.

This book contains seven chapters that discuss AI and VAs. Examples and scientific detail support the presented information. The chapters have been drafted to provide enough technical information for both general and professional readers.

Chapter 1 introduces VAs and includes a review of the scientific background of VA development from 1910 to the present. This chapter contains a detail of interaction methods in VA technology and related services. It also discusses the ethical implications of VA technology and compares notable VAs available in the market and discusses their economic relevance for individuals and enterprises. The chapter addresses security concerns and clarifies the definition of virtual human assistants.

Chapter 2 is about generating the voice of the interactive virtual assistant (IVA). It presents an overview of the current approaches for generating spoken content using text-to-speech synthesis (TTS) systems and thus the voice of an IVA. The overview builds upon the issues that make spoken content generation a non-trivial task and introduces the two main components of a TTS system: text processing and acoustic modeling. It then focuses on providing the reader with the minimally required scientific details of the terminology and methods involved in speech synthesis, yet with sufficient knowledge to make the initial decisions regarding the choice of technology for the vocal identity of the IVA.

In Chapter 3, an IT manager from IBM discusses the use of VAs and related ethical challenges. VAs are becoming a part of our daily lives, both in our homes and our workplaces. Sometimes we may not even know that the customer service agent we are speaking with is a VA. These assistants continuously collect information from our interactions and learn many things about us. The information they gather over time is enormous. This chapter introduces the concept of ethics and discusses the ethical principles of VAs (transparency, justice and fairness, non-maleficence, responsibility, and privacy). Although there is limited regulation governing VAs, practical guidelines and recommendations are provided for designers and developers to understand the ethical implications when building a VA. The chapter also discusses technology and learning techniques for VAs and presents examples of how to ensure they are ethical.

Chapter 4 covers type-theoretic human-AI collaboration. In today's information age, we work under the constant drive to be more productive, and we certainly progress towards being an AI-augmented workforce where each of us is augmented by VAs and work together with each other (and their AI assistants) at scale. To achieve this, a framework should facilitate communication across a network of different humans and machines. As advancements in AI (narrow or general) models continue, we will invariably reach a stage where humans and AI co-exist in an interactive and personalized manner, which is different from today's largely invisible AI that mainly operates autonomously in the background. In this chapter, the authors discuss their proposed framework designed to collaborate within a network of humans and VAs. To collaborate, we need a language and a framework. In the context of humans, a human language suffices to describe and orchestrate our intents with others. This, however, is insufficient in the context of humans and machines. Therefore, this framework is built upon type theory (a branch of symbolic logic in mathematics), enabling the type of theoretic description, composition, and orchestration of intent and implementations for an AI-augmented workforce.

In Chapter 5, a research team from the University of L'Aquila, Italy discusses AI-powered VAs in banking and financial services. The chapter provides a framework for analysis of evolutionary trends in finance that have to do with technological progress, especially with AI applications. The starting point can be identified with a survey on how AI has modified the business areas involving banking and financial services and on what can be expected, in terms of future strategic shifts and behavioral changes, on both the supply and demand sides. The next step revolves around a wider and deeper investigation into the role that VAs have started to, and are likely to further, play in the areas under scrutiny. Special attention is paid to the provision of enhanced customer service support, including conversational AI and sound branding.

Chapter 6 presents the specific wear rate (SWR) modeling of polytetrafluoroethylene (PTFE) composites via Artificial Neural Network (ANN) and Adaptive Neuro-Fuzzy Inference System (ANFIS) tools. The ANN and ANFIS models have been recognized as potential and good tools for mathematical modeling of composite materials' complex and nonlinear behavior of SWR. This study examines the modeling and prediction of SWRs of PTFE composites using the ANFIS model. In addition, it compares the performances of the models with the conventional multilinear regression model.

Chapter 7 examines a critical challenge in digital banking: phishing attacks. A phishing attack is one of the most common forms of cybercrime worldwide.

In recent years, phishing attacks have continued to escalate in severity, frequency, and impact. Globally, the attacks cause billions of dollars of losses each year. Cybercriminals use phishing for various illicit activities such as personal identity theft and fraud and to perpetrate sophisticated corporate-level attacks against financial institutions, healthcare providers, government agencies, and businesses. Several solutions using various methodologies have been proposed in the literature to counter web-phishing threats. This chapter adopts a novel strategy for detecting and preventing website phishing attacks, with practical implementation via a browser toolbar add-in.

This book gives readers a better vision of the application of VAs in the digital era. The authors hope that this volume will be a valuable resource for individuals and companies interested in using new technologies to improve their personal and professional lives. The chapters in this volume present the state of the art of VAs and AI. The breadth and depth of coverage make this volume a useful resource for researchers in academia and industrial specialists. The editor hopes that this book will spur further discussions on using VA technology in different industries.

Ali Soofastaei
Artificial Intelligence Center,
Vale,
Brisbane, Australia

Chapter 1

Introductory Chapter: Virtual Assistants

Ali Soofastaei

1. Introduction

The application of Virtual Assistants (VAs) is growing fast in our personal and professional life. It has been predicted that 25% of households using a VA will have two or more devices by 2021 [1]. A virtual assistant is an intelligent application that can perform tasks or provide services for a person responding to orders or inquiries. Some VAs can understand and respond to human speech using synthesized voices. Users may use voice commands to request their VA to answer the questions, manage home appliances, control media playing, and handle other essential activities like email, creating the actions lists, and organize the meetings on calendars [2]. In the Internet of Things (IoT) world, an VA is a popular service to communicate with users based on voice command.

VA capabilities and usage are rapidly rising, thanks to new technologies reaching the people's requirements and a robust focus on voice user interfaces. Samsung, Google, and Apple each have a considerable smartphone user base. Microsoft's Windows-based personal computers, smartphones, and smart speakers have an intelligent VA installed base. On Amazon, smart speakers have a sizable installed base [3]. Over 100 million people have used Conversica's short message and email interface Intelligent Virtual Assistants (IVAs) services in their companies.

Famous virtual assistants like Amazon Alexa and Google Assistant are typically cloud-based for maximum performance and data management. Many behavioral traces, including the user's voice activity history with extensive descriptions, can be saved in a VA ecosystem's remote cloud servers during this process.

The VAs story started in the 1910s, and the growth of technology has supported VAs' improvement. The application of Artificial Intelligence (AI) also was a turning point in VAs journey. Using AI to develop the VAs was a great jump to increase the VAs' capabilities. Currently, VAs use narrow AI with limited options. However, using general AI in the near future can be a revolution to improve the quality of VAs' services.

2. Backgrounds

2.1 Investigational years: 1910s: 1980s

In 1922, an interesting toy named Radio Rex was introduced that was the first voice-activated doll [4]. A toy in the dog shape would appear from its den the moment it was given a name.

Bell Labs introduced the "Audrey," which was an Automatic Digit Identification device in 1952. It took up a six-foot-high relay rack, used much power, had many wires, and had all of the issues that come with complicated vacuum-tube electronics. Despite this, Audrey was able to discriminate between phonemes, which are the basic components of speech. However, it was restricted to precise digit identification by assigned speakers. As a result, it may be utilized for voice dialing. However, push-button dialing was generally less expensive and faster than pronouncing the digits in order [5].

Another early gadget that could carry out digital language identification was Shoebox voice-activated calculator that IBM developed. It was revealed to the public for the period of the 1962 Seattle World's Fair after its first market debut in 1961. This initial machine, which was built nearly twenty years earlier than the first Personal Computer made by IBM and debuted in 1981, was capable of detecting sixteen verbal phrases and the numbers 0 through 9.

ELIZA, the first Natural Language Processing (NLP) application or chatbot, was invented by MIT in the 1960s. ELIZA was designed in order to "show that man-machine interaction is essentially superficial" [6]. It applied configuration matching and replacement procedures in written reactions to simulate conversation, creating the impression that the machine understood what was being said.

The ELIZA was designed by professor Joseph Weizenbaum. During the ELIZA development period, Joseph's assistant has requested that he leave the room so that she and ELIZA can chat. Professor Weizenbaum later remarked, "I had no idea that brief exposures to a really simple computer software might cause serious delusional thinking in otherwise normal people [7]." The ELIZA impact, or the tendency to instinctively believe machine activities are equal to people's behaviors, was called after this. Anthropomorphizing is a phenomenon that occurs in human interactions with VAs.

When DARPA funded a five-year Speech Understanding Research effort at Carnegie Mellon in the 1970s, the goal was to reach a vocabulary of 1,000 words. Participants included IBM, Carnegie Mellon University (CMU), and Stanford Research Institute, among many others.

The result was "Harpy," a robot that could understand speech and knew around 1000 words, roughly equivalent to a three-year vocabulary. To reduce voice recognition failures, it could also analyze speech that followed pre-programmed vocabularies, pronunciations, and grammatical patterns to determine which word sequences made sense when spoken.

An improvement to the Shoebox was released in 1986 with the Tangora, a speech recognition typewriter. With a vocabulary of 20,000 words, it was able to anticipate the most likely outcome based on its information. Because of this, it was given the name "Fastest Typewriter. As part of its digital signal processing, IBM used a Hidden Markov model, which integrates statistics into the Using this strategy, you may anticipate which phonemes will follow a given phoneme. However, every speaker was responsible for training the typewriter to recognize his or her voice and halt in.

2.2 The beginning of intelligent virtual assistants: 1990s: Present

To compete for customers in the 1990s, companies such as IBM, Philips, and Lemont & Hauspie began integrating digital voice recognition into personal computers. The first smartphone introduced in 1994, the IBM Simon laid the groundwork for today's smart virtual assistant.

In 1997, Dragon's Biologically Talking application was able to detect and transcribe natural human speech at a pace of 100 words per minute, with no gaps between syllables. Biologically Talking is still accessible for download, and many doctors in the United States and the United Kingdom continue to use it to keep track of their medical records.

In 2001, Colloquies released Smarter Child on AIM and MSN Messenger, among other platforms. "Smarter Child" can play games and check the weather as well as seek up data. It can even speak with others to a certain extent, even if it is text.

Siri, which debuted on October 4, 2011, as an option of the iPhone 4S, was the first innovative digital VA to be placed on a smartphone [8, 9]. Siri was built when Apple Inc. purchased Siri Inc. in 2010, a spin-off of SRI International, a research institute financed by DARPA and the US Department of Defense [10]. It was created to make texting, making phone calls, checking the weather, and setting the alarm easier. In addition, it can now make restaurant recommendations, perform Internet searches, and offer driving directions.

Amazon debuted Alexa alongside the Echo in November 2014. Later, in April 2017, Amazon launched a facility that allows users to create conversational interfaces for any VA or interface.

From 2017 till 2021, all the VAs mentioned above have been developed, and there are the more intelligent VAs using for individuals and professional activities. The companies in different areas use the VAs to improve the quality of their decisions at different levels, from operation to the high management level.

3. Virtual assistants - method of interaction

There are different methods of interaction that VAs are using them. In the following, three of the popular VAs' interaction methods are mentioned.

- Text, including online chat, text messages, email, as well as other text-based modes of interaction, for instance, Conversica's IVAs for enterprise [11].

- Voice, for instance, with Siri on an iPhone (Apple products), Google Assistant on Android mobile smartphones, or Amazon Alexa [12] on the Amazon Echo device.

- By shooting and uploading photos, as Bixby on the Samsung Galaxy does.

Various VAs, such as Google Assistant, are available in several ways, including chat on the Google Allo and Google Messages apps and voice on Google Home smart speakers.

VAs use NLP to translate text input from the user or voice input into executable. Furthermore, many people use AI techniques, such as machine learning, to learn continuously. Some of these assistants, such as Google Assistant (which includes Google Lens) and Samsung Bixby, can also perform image processing to distinguish things in the image, allowing users to obtain better results from the images they have clicked.

The awake phrase could be used to activate a voice-activated assistant. These words, for example, are "OK Google," or "Hey Google," "Hey Siri," "Alexa," and "Hey Microsoft" [13]. However, there are increasing legal dangers associated with VAs as they become more popular [14].

4. Virtual assistants: Services

VAs can help with a wide range of tasks. These include the following [15].

- Set the alarm, construct to-do and shopping lists, and support data such as weather;
- Play TV shows, serials, and films on TVs, running from, e.g., Amazon Prime, YouTube;
- Play music from the platforms such as Pandora, YouTube, Spotify, and; podcasts, and read journals and audiobooks;
- Assist citizens in their dealings with the government;
- Conversational business; and
- Humans should be used to supplement and replace customer service [16]. For example, one study indicated that an automated online assistant reduced the burden of a human-staffed call center by 30% [17].

5. Virtual assistants - ethics implications

Generally, the consumers provide free data for the preparation and development of AI algorithms and VAs, which is often ethically disturbing. However, knowing how the applied intelligent models are developed using the consumers' data and information could be more ethically troubling.

Most VAs on the market use Artificial Neural Networks (ANNs) to train AI algorithms, requiring many labeled data. In order to understand the increase in microwork over the past decade, however, this information must be categorized by a human being in order to understand the increase in microwork over the past decade. However, a human being must categorize this information. People worldwide are paid to perform repetitive and incredibly simple tasks, such as listening and copying down voice input from a virtual assistant for a few cents. Because of the insecurity it creates and the lack of control, microwork has been called out as a problem. The average hourly wage was 1.38 dollars [18], with no healthcare, sick pay, retirement benefits, or minimum salary. This has led to a dispute between VAs and their designers over employment insecurity, and the AIs they propose are still human in a way that would be impossible without millions of human workers micromanaging them [19].

A VA provider's unencrypted access to voice commands raises privacy concerns since they can be shared with third parties and handled unlawfully or unexpectedly [20]. Along with language content, a user's style of expression and voice features can provide information about his or her biometric identification, personality attributes, physical and mental health condition, sex and gender, moods and emotions, and socioeconomic status and geographic origin [21].

6. Comparison of notable virtual assistants

Different AI products work as VA in the market. Each product has been designed to provide the assistant service for the specific product. There also are different brands of VAs, and behind them are genius companies who annually are investing billion dollars in this field. **Table 1** shows a shortlist of the most used VAs and their capabilities.

Virtual Assistant	Developer	IOT	Chromecast Integration	Smart Phone App
Alexa	Amazon	Yes	No	Yes
Alice	Yandex	Yes	No	Yes
AliGenie	Alibaba	Yes	No	Yes
Assistant	Speaktoit	No	No	Yes
Bixby	Samsung	No	No	Yes
BlackBerry Assistant	BlackBerry	No	No	Yes
Braina	Brain soft	No	No	Yes
Clova	Naver	Yes	No	Yes
Cortana	Microsoft	Yes	No	Yes
Duer	Baidu	N/A	N/A	N/A
Evi	Amazon	No	No	Yes
Google Assistant	Google	Yes	Yes	Yes
Google Now	Google	Yes	Yes	Yes
M	Facebook	N/A	N/A	N/A
Mycroft	Mycroft	Yes	Yes	Yes
SILVIA	Cognitive Code	No	No	Yes
Siri	Apple Inc.	Yes	No	Yes
Viv	Samsung	Yes	No	Yes
Xiaowei	Tencent	N/A	N/A	N/A
Celia	Huawei	Yes	No	Yes

Table 1.
Notable virtual assistants.

7. Virtual assistants – Financial importance

7.1 For persons

Digital experiences facilitated by VAs are one of the most encouraging end-user trends in recent years. Specialists predict that digital practices would gain a prestige equivalent to 'real' ones, if not more sought-after and valued [22]. The development is supported by frequent users and a significant increase in the number of virtual digital assistant users worldwide. The number of people who use digital VAs regularly was predicted to be approximately 1 billion in mid-2017 [23]. Furthermore, virtual digital assistant technology is no longer limited to smartphone apps but is also found in many different industries [24]. There will be a 34.9 percent CAGR for speech recognition technology from 2016 to 2024, surpassing a global market size of US$7.5 billion by 2024 [25] as a result of considerable R&D expenditures of enterprises across all sectors and increasing use of mobile devices in speech recognition technology [24].

According to an Ovum estimate, by 2021, the "native digital assistant installed base" will outnumber the global population, with 7.5 billion active speech AI-capable devices [25]. "Google Assistant would dominate the speech AI–capable device market with 23.3 percent market share by that time," according to Ovum, "followed by Samsung's Bixby (14.5 percent), Apple's Siri (13.1 percent), Amazon's Alexa (3.9 percent), and Microsoft's Cortana (2.3 percent)" [25].

Businesses in North America (such as Nuance Communications and IBM) are projected to dominate the sector over the next few years due to BYOD (Bring Your Own Device) and enterprise mobility business strategies. Furthermore, the growing demand for smartphone-assisted platforms will likely propel IVA's further growth in North America. On the other hand, even though it is smaller than the North American market [24], the intelligent VA sector in the Asia-Pacific area is predicted to expand at a 40 percent annual growth rate (above the world average) between 2016 and 2024, with its primary players located in India and China.

7.2 For companies

VAs should not be viewed solely as a tool for individuals, as they may have genuine economic value for businesses. A virtual assistant, for example, can serve as an always-available aide with encyclopedia knowledge. Furthermore, it can organize meetings, checking inventories, and verifying data. VAs, on the other hand, are so significant that their integration into small and medium-sized businesses is frequently a simple first step towards a more worldwide adaption and use of IoT. Small and medium-sized enterprises regard IoT technologies as critical technologies that are difficult, risky, or expensive to employ [26].

8. Virtual assistants: Protection

To demonstrate how audio commands can be directly integrated into music or spoken text, researchers from the University of California, Berkeley, published a study in May 2018. The publication showed that VAs could perform specified actions without the user's knowledge. The researchers altered audio files to eliminate the sound patterns that speech recognition algorithms are designed to recognize. Instead, noises were used to direct the system to dial numbers on the phone, launch webpages, or even move money [27]. Since 2016 [27], this has been a possibility, and it impacts Apple, Amazon, and Google devices [28].

Security and privacy concerns with IVAs are not limited to unwanted actions or voice recording. For example, when a person pretends to be someone else, he or she uses malevolent voice commands to gain illegal access to their home or garage, such as unlocking a smart door or shopping items online without their The system may have trouble distinguishing between similar sounds, even though some IVAs have a voice-training feature to prevent imitating. In addition, the system may be fooled into believing that the user is the real owner if a malicious individual has access [29] to an IVA-enabled device.

9. Virtual assistants: A new definition

From 2020 when the world faced the COVID-19 pandemic, and most people had to work remotely from home, the VA found the new definition. A human VA, also known as a virtual office assistant, is a self-employed person who works remotely from a home office to give clients professional administrative, technical, or creative (social) help [30]. Unless these indirect costs are included in the VA fees, clients are not liable for employee-related taxes and insurance or benefits because VAs are independent contractors rather than employees. They also avoid the logistical nightmare of providing additional office space, equipment, or supplies to a third party. A virtual assistant (VA) is a person who does a specific task for a client. The client pays only for work, that is. VAs typically serve other small companies [31, 32], but they can also assist busy executives.

VAs use the Internet, email, phone-call conferences and online workplaces, as well as fax machines, to communicate and exchange data with each other There is also a growing use of Skype and Zoom, Slack, or Google Voice by virtual assistants (VAs Because the professionals in this field operate on contract, it is believed that they will work together for a long Ten years of office experience is required for executive assistants, office managers/supervisors, secretaries, paralegal assistants, legal secretaries, real estate assistants, and information technology.

Voice Over Internet Protocol (VOIP) services like Skype, Microsoft Teams, Google Meet, and Zoom have made it possible to have a virtual assistant (VA) answer the phone without the end user's knowledge in recent years, and VAs have made their way into many mainstream organizations. With today's technology, many firms may personalize their receptionists without having to pay for an additional receptionist.

A VA is a person or company who works remotely as an independent professional, supplying a wide range of products to businesses and customers. When it comes to the typical secretarial tasks such as website editing, social media marketing, and customer care, and data input and accounting (MYOB and QuickBooks), virtual assistants excel. In the virtual world, the industry has changed tremendously as more people join.

VAs come from various professional backgrounds, but the majority have at least a few years of experience working in the "real" (non-virtual) corporate sector or working online or remotely. The modern world is a place for VAs, where the next generation is increasingly relying on intelligent technology to improve their personal and professional lives.

Author details

Ali Soofastaei
Artificial Intelligence Center, Vale, Brisbane, Australia

*Address all correspondence to: ali@soofastaei.net

References

[1] IDC. (2015). Explosive Internet of Things Spending to Reach$1.7 trillion in 2020 According to IDC. http://www.businesswire.com/news/home/20150602005329/en/

[2] Klüwer, Tina. "From chatbots to dialog systems." Conversational agents and natural language interaction: Techniques and Effective Practices. IGI Global, 2011. 1-22.

[3] Daniel B. Kline (January 30, 2017). "Alexa, How Big Is Amazon's Echo"Alexa, How Big Is Amazon's Echo?". The Motley Fool.

[4] Markowitz, Judith. "Toys That Have a Voice." Speech Tech Mag.

[5] Moskvitch, Katia. "The machines that learned to listen." BBC.com. Retrieved May 5, 2020.

[6] Epstein, J; Klinkenberg, W. D (1 May 2001). "From Eliza to Internet: a brief history of computerized assessment." Computers in Human Behaviour. 17 (3): 295-314. ISSN 0747-5632.

[7] Weizenbaum, Joseph (1976). Computer power and human reason: from judgment to calculation. Oliver Wendell Holmes Library Phillips Academy. San Francisco: W. H. Freeman.

[8] "Smartphone: your new personal assistant – Orange Pop." July 10, 2017. Archived from the original on July 10, 2017. Retrieved May 5, 2020.

[9] Darren Murph (October 4, 2011). "iPhone 4S hands-on!". Retrieved December 10, 2017.

[10] "Feature: Von IBM Shoebox bis Siri: 50 Jahre Spracher kennung – WELT" [From IBM Shoebox to Siri: 50 years of speech recognition]. Die Welt (in German). Welt.de. April 20, 2012. Retrieved December 10, 2017.

[11] "Conversica Raises $31 Million in Series C Funding to Fuel Expansion of Conversational AI for Business". Bloomberg.com. October 30, 2018. Retrieved October 23, 2020.

[12] Herrera, Sebastian. "Amazon Extends Alexa's Reach into Wearables." The Wall Street Journal. Retrieved September 26, 2019.

[13] "S7617 – Developing Your Wake Word Engine Just Like 'Alexa' and 'OK Google". GPU Technology Conference. Retrieved July 17, 2017.

[14] Van Loo, Rory (March 1, 2019). "Digital Market Perfection." Michigan Law Review. 117 (5): 815.

[15] Taylor Martin; David Priest (September 10, 2017). "The complete list of Alexa commands so far." CNET. Retrieved December 10, 2017.

[16] Kongthon, Alisa; Sangkeettrakarn, Chat chawal; Kong young, Sara woot; Haruechaiyasak, Choo chart (1 January 2009). Implementing an Online Help Desk System Based on Conversational Agent. Proceedings of the International Conference on Management of Emergent Digital EcoSystems. MEDES '09. New York, NY, USA: ACM. pp. 69:450-69:451.

[17] Anthony O'Donnell (June 3, 2010). "Aetna's new "virtual online assistant." Insurance & Technology. Archived from the original on June 7, 2010.

[18] Horton, John Joseph; Chilton, Lydia B. (2010). "The labor economics of paid crowdsourcing." Proceedings of the 11th ACM Conference on Electronic Commerce – EC '10. New York, New York, USA: ACM Press: 209.

[19] Casilli, Antonio A. (2019). En attendant les robots. Enquête sur le travail du clic. Editions Seuil. ISBN 978-2-02-140188-2.

[20] Apple, Google, and Amazon May Have Violated Your Privacy by Reviewing Digital Assistant Commands". Fortune. August 5, 2019. Retrieved May 13, 2020.

[21] Kröger, Jacob Leon; Lutz, Otto Hans-Martin; Reschke, Philip (2020). "Privacy Implications of Voice and Speech Analysis – Information Disclosure by Inference." 576: 242-258.

[22] "5 Consumer Trends for 2017". Trend Watching. October 31, 2016. Retrieved December 10, 2017.

[23] Felix Richter (August 26, 2016). "Chart: Digital Assistants – Always at Your Service." Statista. Retrieved December 10, 2017.

[24] "Virtual Assistant Industry Statistics « Global Market Insights, Inc." Gminsights.wordpress.com. January 30, 2017. Retrieved December 10, 2017.

[25] "Virtual digital assistants to overtake world population by 2021". ovum.informa.com. Retrieved May 11, 2018.

[26] Jones, Nory B.; Graham, C. Matt (February 2018). "Can the IoT Help Small Businesses?". Bulletin of Science, Technology & Society. 38 (1-2): 3-12.

[27] "Alexa and Siri Can Hear This Hidden Command. You Cannot". The New York Times. May 10, 2018. ISSN 0362-4331. Retrieved May 11, 2018.

[28] "As voice assistants go mainstream, researchers warn of vulnerabilities." CNET. May 10, 2018. Retrieved May 11, 2018.

[29] Chung, H.; Iorga, M.; Voas, J.; Lee, S. (2017). "Alexa, Can I Trust You?". Computer. 50 (9): 100-104.

[30] Starks, Misty (July–August 2006). "Helping Entrepreneurs, virtually" (PDF). D-MARS. Archived from the original (PDF) on 2008-09-21. Retrieved 2008-07-27.

[31] Finkelstein, Brad (February–March 2005). "Virtual Assistants A Reality." Broker Magazine. 7 (1): 44-46.

[32] Johnson, Tory (2007-07-23). "Work-From-Home Tips: Job Opportunities for Everyone." ABC News. Retrieved 2008-07-28.

Chapter 2

Generating the Voice of the Interactive Virtual Assistant

Adriana Stan and Beáta Lőrincz

Abstract

This chapter introduces an overview of the current approaches for generating spoken content using text-to-speech synthesis (TTS) systems, and thus the voice of an Interactive Virtual Assistant (IVA). The overview builds upon the issues which make spoken content generation a non-trivial task, and introduces the two main components of a TTS system: text processing and acoustic modelling. It then focuses on providing the reader with the minimally required scientific details of the terminology and methods involved in speech synthesis, yet with sufficient knowledge so as to be able to make the initial decisions regarding the choice of technology for the vocal identity of the IVA. The speech synthesis methodologies' description begins with the basic, easy to run, low-requirement rule-based synthesis, and ends up within the state-of-the-art deep learning landscape. To bring this extremely complex and extensive research field closer to commercial deployment, an extensive indexing of the readily and freely available resources and tools required to build a TTS system is provided. Quality evaluation methods and open research problems are, as well, highlighted at end of the chapter.

Keywords: text-to-speech synthesis, text processing, deep learning, interactive virtual assistant

1. Introduction

Generating the voice of an interactive virtual assistant (IVA) is performed by the so called *text-to-speech synthesis (TTS)* systems. A TTS system takes raw text as input and converts it into an acoustic signal or waveform, through a series of intermediate steps. The synthesised speech commonly pertains to a single, pre-defined speaker, and should be as natural and as intelligible as human speech. An overview of the main components of a TTS system is shown in **Figure 1**.

At first sight this seems like a straightforward mapping of each character in the input text to its acoustic realisation. However, there are numerous technical issues

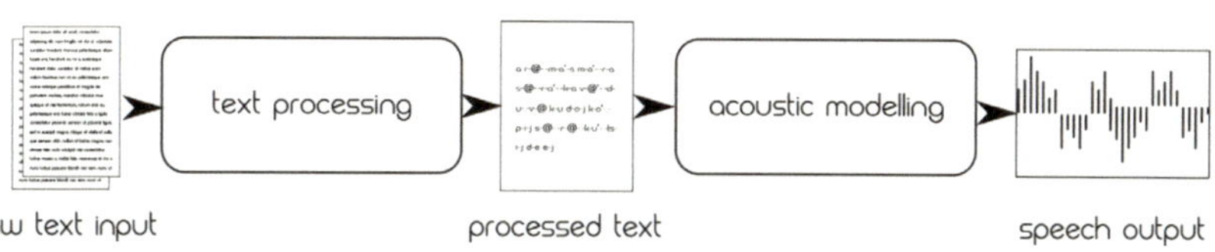

Figure 1.
Overview of a text-to-speech synthesis system's main components.

which make natural speech synthesis an extremely complex problem, with some of the most important ones being indexed below:

the written language is a discrete, compressed representation of the spoken language aimed at transferring a message, irrespective of other factors pertaining to the speaker's identity, emotional state, etc. Also, in almost any language, the written symbols are not truly informative of their pronunciation, with the most notable example being English. The pronunciation of a letter or sequence of letters which yield a single sound is called a *phone*. One exception here is the Korean alphabet for which the symbols approximate the position of the articulator organs, and was introduced in 1443 by King Sejong the Great to increase the literacy among the Korean population. But for most languages, the so called orthographic transparency is rather opaque;
the human ear is highly adapted to the frequency regions in which the relevant information from speech resides (i.e. 50–8000 Hz). Any slight changes to what is considered to be natural speech, any artefacts, or unnatural sequences present in a waveform deemed to contain spoken content, will be immediately detected by the listener;
speaker and speech variability is a result of the uniqueness of each individual. This means that there are no two persons having the same voice timbre or pronouncing the same word in a similar manner. Even more so, one person will never utter a word or a fixed message in an exactly identical manner even when the repetitions are consecutive;
co-articulation effects derive from the articulator organs' inertial movement. There are no abrupt transitions between sounds and, with very few exceptions, it is very hard to determine the exact boundary of each sound. Another result of the co-articulation is the presence of reductions or modifications in the spoken form of a word or sequence of words, derived from the impossibility or hardship of uttering a smooth transition between some particular phone pairs;
prosody is defined as the rhythm and melody or intonation of an utterance. The prosody is again related to the speaker's individuality, cultural heritage, education and emotional state. There are no clear systems which describe the prosody of a spoken message, and one's person understanding of, for example, portraying an angry state of mind is completely different from another;
no fixed set of measurable factors define a speaker's identity and speaking characteristics. Therefore, when wanting to reproduce one's voice the only way to do this for now is to record that person and extract statistical information from the acoustic signal;
no objective measure correlates the physical representation of a speech signal with the perceptual evaluation of a synthesised speech's quality and/or appropriateness.

The problems listed above have been solved, to some extent, in TTS systems by employing high-level machine learning algorithms, developing large expert resources or by limiting the applicability and use-case scenarios for the synthesised speech. In the following sections we describe each of the main components of a TTS system, with an emphasis on the acoustic modelling part which poses the greatest problems as of yet. We also index some of the freely available resources and tools which can aid a fast development of a synthesis system for commercial IVAs in a dedicated section of the chapter, and conclude with the discussion of some open problems in the final section.

2. Speech processing fundamentals

Before diving into the text-to-speech synthesis components, it is important to define a basic set of terms related to digital speech processing. A complete overview of this domain is beyond the scope of this chapter, and we shall only refer to the terms used to describe the systems in the following sections.

Speech is the result of the air exhaled from the lungs modulated by the articulator organs and their instantaneous or transitioning position: vocal cords, larynx, pharynx, oral cavity, palate, tongue, teeth, jaw, lips and nasal cavity. By modulation we refer to the changes suffered by the air stream as it encounters these organs. One of the most important organs in speech are the vocal cords, as they determine the periodicity of the speech signal by quickly opening and closing as the air passes through. The vocal cords are used in the generation of vowels and voiced consonant sounds [1]. The perceived result of this periodicity is called the *pitch*, and its objective measure is called *fundamental frequency*, commonly abbreviated F_0 [2]. The slight difference between pitch and F_0 is better explained by the auditory illusion of the *missing fundamental* [3] where the measured fundamental frequency differs from the perceived pitch. Commonly, the terms are used interchangeably, but readers should be aware of this small difference. The pitch variation over time in the speech signal gives the melody or intonation of the spoken content. Another important definition is that of *vocal tract* which refers to all articulators positioned above the vocal cords. The resonance frequencies of the vocal tract are called *formant frequencies*. Three formants are commonly measured and noted as F_1, F_2 and F_3.

Looking into the time domain, as a result of the articulator movement, the speech signal is not stationary, and its characteristics evolve through time. The smallest time interval in which the speech signal is considered to be *quasi-stationary* is 20–40 msec. This interval determines the so-called *frame-level analysis* or *windowing* of the speech signal, in which the signal is segmented and analysed at more granular time scales for the resulting analysis to adhere to the digital signal processing theorems and fundamentals [4].

The *spectrum* or *instantaneous spectrum* is the result of decomposing the speech signal into its frequency components through Fourier analysis [5] on a frame-by-frame basis. Visualising the evolution of the spectrum through time yields the *spectrogram*. Because the human ear has a non-linear frequency response, the linear spectrum is commonly transformed into the *Mel spectrum*, where the Mel frequencies are a non-linear transformation of the frequency domain pertaining to the pitches judged by listeners to be equal in distance one from another. Frequency domain analysis is omnipresent in all speech related applications, and Mel spectrograms are the most common representations of the speech signal in the neural network-based synthesis.

One other frequency-derived representation of the speech is the *cepstral* [6] representation which is a transform of the spectrum aimed at separating the vocal tract and the vocal cord (or glottal) contributions from the speech signal. It is based on homomorphic and decorrelation operations.

3. Text processing

Text processing or *front-end processing* represents the mechanism of generating supplemental information from the raw input text. This information should yield a representation which is hypothetically closer and more relevant to the acoustic realisation of the text, and therefore tightens the gap between the two domains. Depending on the targeted language, this task is more or less complex [2]. A list of the common front-end processing steps is given below:

text tokenisation splits the input text into syntactically meaningful chunks i.e. phrases sentences and words. Languages which do not have a word separator such as Chinese or Japanese pose additional complexity for this task [7];
diacritic restoration - in languages with diacritic symbols it might be the case that the user does not type these symbols and this leads to an incorrect spoken sequence [8]. The diacritic restoration refers to adding the diacritic symbols back into the text so that the intended meaning is preserved;

text normalisation converts written expressions into their '"spoken" forms e.g. $3.16 is converted into "three dollars sixteen cents." or 911 is converted into "nine one one" and not "nine hundred eleven" [9]. An additional problem is caused by languages which have genders assigned to nouns e.g. in Romanian "21 oi = douăzeci şi *una* de oi" (en. twenty one sheep–feminine) versus "21 cai = douăzeci şi *unu* de cai" (en. twenty one horses-masculine);

part-of-speech tagging (POS) assigns a part-of-speech (i.e. noun, verb, adverb, adjective, etc.) to each word in the input sequence. The POS is important to disambiguate non-homophone homographs. These are words which are spelled the same but pronounced differently based on their POS (e.g. *bow* - to bend down/the front of a boat/tied loops). POS are also essential for placing the accent or focus of an utterance on the correct word or word sequence [10];

lexical stress marking - the lexical stress pertains to the syllable within a word which is more prominent [11]. There are however languages for which this notion is quite elusive such as French or Spanish. Yet in English a stress-timed language assigning the correct stress to each word is essential for conveying the correct message. Along with the POS the lexical stress also helps disambiguate non-homophone homographs in the spoken content. There are also phoneticians who would mark a secondary and tertiary stress but for speech synthesis the primary stress should be enough as the secondary does not affect the meaning but rather the naturalness or emphasis of the speech;

syllabification - syllables represent the base unit of co-articulation and determine the rhythm of speech [12]. Again different languages pose different problems and languages such as Japanese rely on syllables for their alphabetic inventory. As a general rule every syllable has only one vowel sound but can be accompanied by semi-vowels. Compound words generally do not follow the general rules such that prefixes and suffixes will be pronounced as a single syllable;

phonetic transcription is the final result of all the steps above. Meaning that by knowing the POS the lexical stress and syllabification of a word the exact pronunciation can be derived [13]. The phones are a set of symbols corresponding to an individual articulatory target position in a language or otherwise put it is the fixed sound alphabet of a language. This alphabet determines how each sequence of letters should be pronounced. Yet this is not always the case and the concept of orthographic transparency determines the ease with which a reader can utter a written text in a particular language;

prosodic labels, phrase breaks - with all the lexical information in place there is still the issue of emphasising the correct words as per intent of the writer. The accent and pauses in speech are very important and can make the message decoding a very complex task or an easier one with the information being able to be faster assimilated by the listener. There is quite a lot of debate on how the prosody should be marked in text and if it should be [14]. There is definitely some markings in the form of punctuation signs yet there is a huge gap between the text and the spoken output. However public speaking coaching puts a large weight on the prosodic aspect of the speech and therefore captivating the listeners attention through non-verbal queues;

word/character embeddings - are the result of converting the words or characters in the text into a numeric representation which should encompass more information about their identity pronunciation syntax or meaning than the surface form does. Embeddings are learnt from large text corpora and are language dependent. Some of the algorithms used to build such representations are: Word2Vec [15] GloVe [16] ELMo [17] and BERT [18].

4. Acoustic modelling

The acoustic modelling or *back-end processing* part refers to the methods which convert the desired input text sequence into a speech waveform. Some of the earliest proofs of so-called talking heads are mentioned by Aurrilac (1003 A.D.), Albert Magnus (1198–1280) or Roger Bacon (1214–1294). The first electronic synthesiser was the VODER (Voice Operation DEmonstratoR) created by Homer Dudley at Bell Laboratories in 1939. The VODER was able to generate speech by tediously operating a keyboard and foot pedals to control a series of digital filters.

Coming to the more recent developments, and based on the main method of generating the speech signal, speech synthesis systems can be classified into **rule-based** and **corpus-based** methods. In rule-based methods, similar to the VODER, the sound is generated by a fixed, pre-computed set of parameters.

Corpus-based methods, on the other hand, use a set of speech recordings to generate the synthetic output or to derive statistical parameters from the analysis of the spoken content. It can be argued that using pre-recorded samples is not in itself synthesis, but rather a speech collage. In this sense Taylor gives a different definition of speech synthesis: *"the output of a spoken utterance from a resource in which it has not been prior spoken"* [2].

4.1 Rule-based synthesis

Formant synthesis is one of the first digital methods of speech generation. It is still used today, especially by phoneticians who study various spoken language phenomena. The method uses the approximation of several speech parameters (commonly the F_0 and formant frequencies) for each phone in a language, and also how these parameters vary when transitioning from one phone to the next one [19]. The most representative model of formant synthesis is the one described by [20], which later evolved into the commercial system of MITalk [21]. There are around 40 parameters which describe the formants and their respective bandwidths, and also a series of frequencies for nasals or glottal resonators.

The advantages of formant synthesis are related to the good intelligibility even at high speeds, and its very low computation and memory requirements, making it easy to deploy on limited resource devices. The major drawback of this type of synthesis is, of course, its low quality and robotic sound, and also the fact that for high-pitched outputs, the formant tracking mechanisms can fail to determine the correct values.

Articulatory synthesis uses mechanical and acoustic models of speech production [1]. The physiological effects such as the movement of the tongue, lips, jaw, and the dynamics of the vocal tract and glottis are modelled. For example, [22] uses lip opening, glottal area, opening of nasal cavities, constriction of tongue, and rate between expansion and contraction of the vocal tract along with the first four formant frequencies. Magnetic resonance imaging offers some more insight into the muscle movement [23], yet the complexity of this type of synthesis makes it rather unfeasible for high naturalness and commercial deployment. One exception in the project GNUSpeech [24] but its results are still poor compared to what corpus-based synthesis is able to achieve nowadays.

4.2 Corpus-based synthesis

4.2.1 Concatenative synthesis

As the name entails, concantenative synthesis is a method of producing spoken content by concatenating pre-recorded speech samples. In its most basic form, a concatenative synthesis system contains recordings of all the words needed to be uttered, which are then combined in a very limited vocabulary scenario. For example, in a rudimentary IVA, it will combine the typed-in phone number of a customer by combining pre-recorded digits. Of course, in a large vocabulary, open-domain system, pre-recording all the words in a language is unfeasible. The solution to this problem is to find a smaller set of acoustic units which can be then combined into any spoken phrase. Based on the type of segment stored in the recorded database, the concatenative synthesis is either **fixed inventory** – segments in the database have the same length, or **variable inventory or unit selection** – segments have variable length. As the basic acoustic unit of any language is its phone set, a first open-domain fixed inventory concatenative synthesis made use of *diphones* [25, 26]. A diphone is the acoustic unit spanning from the middle of a phone to the middle of the next one in adjoining phone pairs. Although this yields a much larger acoustic inventory, the diphones are a better choice than phones because they can model the co-articulation effects. For a primitive diphone concatenation system, the recorded speech corpus would include a single repetition of all the diphones in a language. More elaborate systems use diphones in different context (e.g. beginning, middle or end of a word) and with different prosodic events (e.g. accent, variable duration etc.). Another type of fixed inventory system is based on the use of *syllables* as the concatenation unit [27–29]. Some theories state that the basic unit of speech is the syllable and, therefore, the co-articulation effects between them is minimum [30], but the speech database is hard to design. The average number of unique syllables in one language is in the order of thousands.

A natural evolution of the fixed inventory synthesis is the variable length inventory, or unit selection [31, 32]. In unit selection, the recorded corpus includes segments as small as half-phones and go up to short common phrases. The speech database is either stored as-is, or as a set of parameters describing the exact acoustic waveform. The speech corpus, therefore, needs to be very accurately annotated with information regarding the exact phonetic content and boundaries, lexical stress, syllabification, lexical focus and prosodic trends or patterns (e.g. questions, exclamation, statements). The combination of the speech units into the output spoken phrase is done in an iterative manner, by selecting the best speech segments which minimise a global cost function [31] composed of: a *target cost* - measuring how well a sequence of units matches the desired output sequence, and a *concatenation cost* - measuring how well a sequence of units will be joined together and thus avoid the majority of the concatenation artefacts.

Although this type of synthesis is almost 30 years old, it is still present in many commercial applications. However, it poses some design problems, such as: the need for a very large manually segmented and annotated speech corpus; the control of prosody is hard to achieve if the corpus does not contain all the prosodic events needed to synthesise the desired output; changing the speaker identity requires the database recording and processing to be started from scratch; and there are quite a lot of concatenation artefacts present in the output speech making it unnatural, but which have, in some cases, been solved by using a hybrid approach [33].

4.2.2 Statistical-parametric synthesis

Because concatenative synthesis is not very flexible in terms of prosody and speaker identity, in 1989 a first model of statistical-parametric synthesis based on Hidden Markov Models (HMMs) was introduced [34]. The model is parametric because it does not use individual stored speech samples, but rather parameterises the waveform. And it is statistical because it describes the extracted parameters using statistics averaged across the same phonetic identity in the training data [35]. However this first approach did not attract the attention of the specialists because of its highly unnatural output. But in 2005, the HMM-based Speech Synthesis System (HTS) [36] solved part of the initial problems, and the method became the main approach in the research community with most of its studies aiming at fast speaker adaptation [37] and expressivity [38]. In HTS, a 3 state HMM models the statistics of the acoustic parameters of the phones present in the training set. The phones are clustered based on their identity, but also on other contextual factors, such as the previous and next phone identity, the number of syllables in the current word, the part-of-speech of the current word, the number of words in the sentence, or the number of sentences in a phrase, etc. This context clustering is commonly performed with the help of decision trees and ensures that the statistics are extracted from a sufficient number of exemplars. At synthesis time, the text is converted in a context aware complex label and drives the selection of the HMM states and their transitions. The modelled parameters are generally derived from the source-filter model of speech production [1]. One of the most common vocoders used in HTS is STRAIGHT [39] and it parameterises the speech waveform into F_0, Mel cepstral and aperiodicity coefficients. A less performant, yet open vocoder is WORLD [40]. A comparison of several vocoders used for statistical parametric speech synthesis is presented in [41].

There are several advantages for the statistical-parametric synthesis, such as: the small footprint necessary to store speech information; automatic clustering of speech information–removes the problems of hand-written rules; generalisation–even if for a certain phoneme context there is not enough training data, the phone will be clustered along with similar parameter characteristics; flexibility–the trained models can be easily adapted to other speakers or voice characteristics with minimum amount of adaptation data. However, the parameter averaging yields the so-called *buzziness* and low speaker similarity of the output speech, and for this reason the HTS system has not truly made its way into the commercial applications.

4.2.3 Neural synthesis

In 1943, McCulloch and Pitts [42] introduced the first computational model for artificial neural networks (ANN). And although the incipient ANNs have been successfully applied in multiple research areas, including TTS [43], their learning power comes from the ability to stack multiple neural layers between the input and output. However, it was not until 2006 that the hardware and algorithmic solutions enabled adding multiple layers and making the learning process stable. In 2006, Geoffrey Hinton and his team published a series of scientific papers [44, 45] showing how a many-layered neural network could be effectively pre-trained one layer at a time. These remarkable results set the trend for all automatic machine learning algorithms in the following years, and are the bases of the **deep neural network (DNN)** research field. Nowadays, there are very few machine learning applications which do not cite the DNNs as attaining the state-of-the-art results and performances.

In text-to-speech synthesis, the progression from HMMs to DNNs was gradual. Some of the first impacting studies are those of Ling et al. [46] and Zen et al. [47]. Both papers substitute parts of the HMM-based architecture, yet model the audio on a frame-by-frame basis, maintaining the statistical-parametric approach, and also use the same contextual factors in the text processing part. The first open source tool to implement the DNN-based statistical-parametric synthesis is Merlin [48]. A comparison of the improvements achieved by the DNNs compared to HMMs is presented in [49]. However, these methods still rely on a time-aligned set of text features and their acoustic realisations, which requires a very good frame-level aligner systems, usually an HMM-based one. Also, the sequential nature of speech is only marginally modelled through the contextual factors and not within the model itself, while the text still needs to be processed with expert linguistic automated tools which are rarely available in non-mainstream languages.

An intermediate system which replaces all the components in a TTS pipeline with neural networks is that of [50], but it does not incorporate a single end-to-end network. The first study which removes the above dependencies, and models the speech synthesis process as a sequence-to-sequence recurrent network-based architecture is that of Wang et al. [51]. The architecture was able to *"synthesise fairly intelligible speech"* and was the precursor of the more elaborate Char2Wav [52] and Tacotron [53] systems. Both Char2Wav and Tacotron model the TTS generation as a two step process: the first one takes the input text string and converts it into a spectrogram, and the second one, also called the *vocoder*, takes the spectrogram and converts it into a waveform, either in a deterministic manner [54], or with the help of a different neural network [55]. These two synthesis systems were also the first to alleviate the need for more elaborate text representations, and derived them as an inherent learning process, setting the first stepping stones towards true end-to-end speech synthesis [56]. However, for phonetically rich languages it is common to train the models on phonetically transcribed text, and also to augment the input text with additional linguistic information such as part-of-speech tags which can enhance the naturalness of the output speech [57, 58].

Starting with the publication of Tacotron, the DNN-based speech synthesis research and development area has seen an enormous interest from both the academia and the commercial sides. Most focus has been granted on generating extremely high quality speech, but also to the reduction of the computational requirements and generation speed–which in the DNN domain is called *inference* speed. A major breakthrough was obtained by the second version of Tacotron, Tacotron 2 [59], which achieved naturalness scores very close to human speech. However, both systems' architectures involve attention-based recurrent auto-regressive processes which make the inference step very slow and prone to instability issues, such as word skipping, deletions or repetitions. Also, the recurrent neural networks (RNNs) are known to have high demands in terms of data availability and training time. So that, the next step in DNN-based TTS was the introduction of CNNs, in systems such as DC-TTS [60], DeepVoice 3 [61], ClariNet [62], or ParaNet [63]. The CNNs enable a much better data and training efficiency and also a much faster inference speed through parallel processing. And also, recently, the research community started to look into ways of replacing the auto-regressive attention-based generation, and incorporated duration prediction models which stabilise the output and enable a much faster parallel inference of the output speech [64, 65].

Inspired by the success of the Transformer network [66] in text processing, TTS systems have adopted this architecture as well. Transformer based models include Transformer-TTS [67], FastSpeech [68], FastSpeech 2 [69], AlignTTS [70], JDI-T [71], MultiSpeech [72], or Reformer-TTS [73]. Transformer-based architectures improve the training time requirements, and are capable of modelling longer term dependencies present in the text and speech data.

As the naturalness of the output synthetic speech became very high-quality, researchers started to look into ways of easily controlling the different factors of the synthetic speech, such as duration or style. The go-to solution for this are the Variational AutoEncoders (VAEs) and their variations, which enable the disentanglement of the latent representations, and thus a better control of the inferred features [74–78]. There were also a few approaches including Generative Adversarial Networks (GANs), such as GAN-TTS [79] or [80], but due to the fact that GANs are known to pose great training problems, this direction was not that much explored in the context of TTS.

A common problem in all generative modelling irrespective of deep learning methodologies, is the fact that the true probability distribution of the training data is not directly learned or accessible. In 2015, Rezende et al. [81] introduced the normalising flows (NFs) concept. NFs estimate the true probability distribution of the data by deriving it from a simple distribution through a series of invertible transforms. The invertible transforms make it easy to project a measured data point into the latent space and find its likelihood, or to sample from the latent space and generate natural sounding output data. For TTS, NFs have just been introduced, yet there are already a number of high-quality systems and implementations available, such as: Flowtron [82], Glow-TTS [83], Flow-TTS [84], or Wave Tacotron [56]. From the generative perspective, this approach seems, at the moment, to be able to encompass all the desired goals of a speech synthesis system, but there are still a number of issues which need to be addressed, such as the inference time and latent space disentanglement and control.

All the above mentioned neural systems only solve the first part of the end-to-end problem, by taking the input text and converting it into a Mel spectrogram, or variations of it. For the spectrogram to be converted into an audio waveform, there is the separate component, called the vocoder. And there are also numerous studies on this topic dealing with the same trade-off issue of quality versus speed [85].

WaveNet [55] was one of the first neural networks designed to generate audio samples and achieved remarkably natural results. It is still the one vocoder to beat when designing new ones. However, its auto-regressive processes make it unfeasible for parallel inference, and several methods have been proposed to improve it, such as FFTNet [86] or Parallel WaveNet [87], but the quality is somewhat affected. Some other neural architectures used in vocoders are, of course, the recurrent networks used in WaveRNN [88] and LPCNet [89], or the adversarial architectures used in MelGAN [90], GELP [91], Parallel WaveGAN [92], VocGAN [93]. Following the trend of normalising flows-based acoustic modelling, flow-based vocoders have also been implemented. Some of the most remarkable being: FlowWaveNet [94], WaveGlow [95], WaveFlow [96], WG-WaveNet [97], EWG (Efficient WaveGlow) [98], MelGlow [99], or SqueezeWave [100].

In light of all these methods available for neural speech synthesis, it is again important to note the trade-offs between the quality of output speech, model sizes, training times, inference speed, computing power requirements and ease of control and adaptability. In the ideal scenario, a TTS system would be able to generate natural speech, at an order of magnitude faster than real-time processing speed, on a limited resource device. However, this goal has not yet been achieved by the current state-of-the-art, and any developer looking into TTS solutions should first determine the exact applicability scenario before implementing any of the above methods. It may be the case that, for example, in a limited vocabulary, non-interactive assistant, a simple formant synthesis system implemented on a dedicated hardware might be more reliable and adequate.

Some aspects which we did not take into account in the above enumeration are the multispeaker, multilingual TTS systems. However, in a commercial setup these are not directly required and can be substituted by independent high-quality systems integrated in a seamless way withing the IVA.

5. Open resources and tools

Deploying any research result into a commercial environment requires at least a baseline functional proof-of-concept from which to start optimising and adapting the system. It is the same in TTS systems, where especially the speech resources, text-processing tools, and system architectures can be at first tested and only then developed and migrated to the live solution. To aid this development, the following table indexes some of the most important resources and tools available for text to speech synthesis systems. This is by no means an exhaustive list, but rather a starting point. The official implementations pertaining to the published studies are marked as such. If no official implementation was found, we relied on our experience and prior work to link an open tool which comes as close as possible to the original publication.

Speech and text datasets and resources
Language Data Consortium (LDC) is a repository and distribution point for various language resources. Link: www.ldc.upenn.edu

The European Language Resources Association (ELRA) is a non-profit organisation whose main mission is to make Language Resources for Human Language Technologies available to the community at large. Link: www.elra.info/en/
META-SHARE [101] is an open and secure network of repositories for sharing and exchanging language data, tools and related web services. Link: www.meta-share.org
OpenSLR is a site devoted to hosting speech and language resources, such as training corpora for speech recognition, and software related mainly to speech recognition. Link: www.openslr.org
LibriVox is a group of worldwide volunteers who read and record public domain texts creating free public domain audiobooks for download. Link: www.librivox.org
Mozilla Common Voice is part of Mozilla's initiative to help teach machines how real people speak. Link: www.commonvoice.mozilla.org/en/datasets
Project Gutenberg is an online library of free eBooks. Link: www.gutenberg.org
LibriTTS [102] is a multi-speaker English corpus of approximately 585 hours of read English speech designed for TTS research. Link: www.openslr.org/60/
The Centre for Speech Technology Voice Cloning Toolkit (VCTK) Corpus includes speech data uttered by 109 native speakers of English with various accents. Each speaker reads out about 400 sentences. Link: www.datashare.is.ed.ac.uk/handle/10283/2950
CMU Wilderness Multilingual Speech Dataset [103] is a speech dataset of aligned sentences and audio for some 700 different languages. It is based on readings of the New Testament. Link: www.github.com/festvox/datasets-CMU_Wilderness
Text processing tools
Festival is a complete TTS system, but it enables the use of its front-end tools independently. It supports several languages and dialects. Link: www.cstr.ed.ac.uk/projects/festival/
CMUSphinx G2P tool is a grapheme-to-phoneme conversion tool based on transformers. Link: www.github.com/cmusphinx/g2p-seq2seq
Multilingual G2P uses the eSpeak tool to generate phonetic transcriptions in multiple languages. Link: www.github.com/jcsilva/multilingual-g2p.
Stanford NLP tools includes various text-processing and knowledge extraction tools for English and other languages. Link: www.nlp.stanford.edu/software/
RecoAPy [104] tool includes an easy to use interface for recording prompted speech, but also a set of models able to perform high accuracy phonetic transcription in 8 languages. Link: www.gitlab.utcluj.ro/sadriana/recoapy
word2vec [15] is a word embedding model that learns vector representations of words that capture semantic and other properties of these words from large amounts of text data. Link: code.google.com/archive/p/word2vec/
GloVE [16] is a word embedding method that learns from the co-occurences of words in text corpus obtaining similar vector representations for words that occur in the same context. Link: www.nlp.stanford.edu/projects/glove/
ELMo [17] obtains contextualized word embeddings that model the semantics and syntax of the word, but can learn different representations for various contexts. Link: www.allennlp.org/elmo
BERT [18] is a Transformer-based model that obtains context dependent word embeddings and can process sentences in parallel. Link: www.github.com/google-research/bert
Speech synthesis systems
eSpeak is a formant-based compact open source software speech synthesiser. Link: www.espeak.sourceforge.net/ [Official]
Festival is an unrestricted commercial and non-commercial use framework for building concatenative and HMM-based TTS systems. Link: www.cstr.ed.ac.uk/projects/festival/ [Official]
MaryTTS [105] is an open-source, multilingual TTS platform written in Java supporting diphone and unit selection synthesis. Link: http://mary.dfki.de/ [Official]

HTS [36] is the most commonly used implementation of the HMM-based speech synthesis. Link: http://hts.sp.nitech.ac.jp/ [Official]
Merlin [48] is a Python implementation of DNN models for statistical parametric speech synthesis. Link: www.github.com/CSTR-Edinburgh/merlin [Official]
IDLAK [106] is a project to build an end-to-end neural parametric TTS system within the Kaldi ASR framework. Link: www.idlak.readthedocs.io/en/latest/ [Official]
DeepVoice [50] follows the structure of HMM-based TTS systems, but replaces all its components with neural networks. Link: www.github.com/israelg99/deepvoice
Char2Wav [52] is an end-to-end neural model trained on characters that can synthesise speech with the SampleRNN vocoder. Link: https://github.com/sotelo/parrot [Official]
Tacotron [53] is one of the most frequently used end-to-end neural synthesis systems based on recurrent neural nets and attention mechanism. Link: www.github.com/keithito/tacotron
VoiceLoop [107] is one of the first neural synthesisers which uses a buffer memory instead of recurrent layers and does not require an audio-to-phone alignment. Link: www.github.com/facebookarchive/loop [Official]
Tacotron 2 [59] is an enhanced version of Tacotron which modifies the attention mechanism and also uses the WaveNet vocoder to generate the output speech. Link: www.github.com/NVIDIA/tacotron2
DeepVoice 3 [61] is a fully convolutional synthesis system that can synthesise speech in a multispeaker scenario. Link: www.github.com/r9y9/deepvoice3_pytorch
DCTTS [60] - Deep Convolutional TTS is a synthesis system that implements a two step synthesis, by first learning a coarse and then a fine-grained representation of the spectrogram. Link: www.github.com/tugstugi/pytorch-dc-tts
ClariNet [62] is the first text-to-wave neural architecture for speech synthesis, which is fully convolutional and enables fast end-to-end training from scratch. Link: www.github.com/ksw0306/ClariNet
Transformer TTS [67] replaces the recurrent structures of Tacotron 2 with attention mechanisms. Link: www.github.com/soobinseo/Transformer-TTS
GAN-TTS [79] is a GAN-based synthesis system that uses a generator to produce speech and multiple discriminators that evaluate the naturalness and text-adequacy of the output. Link: www.github.com/yanggeng1995/GAN-TTS
FastSpeech [68] is a novel feed-forward network based on Transformer which generates the Mel-spectrogram in parallel, and uses a teacher-based length predictor to achieve this parallel generation. Link: www.github.com/xcmyz/FastSpeech
FastSpeech 2 [69] is an enhanced version of FastSpeech where the length predictor teacher network is replaced by conditioning the output on duration, pitch and energy from extracted from the speech waveform at training and their predicted values in inference. Link: www.github.com/ming024/FastSpeech2
AlignTTS [70] is a feed-forward Transformer-based network with a duration predictor which aligns the speech and audio. Link: www.github.com/Deepest-Project/AlignTTS
Mellotron [108] is a multispeaker TTS able to emote emotions by explicitly conditioning on rhythm and continuous pitch contours from an audio signal. Link: www.github.com/NVIDIA/mellotron [Official]
Flowtron [82] is an autoregressive normalising flow-based generative network for TTS, also capable of transferring style from one speaker to another. Link: www.github.com/NVIDIA/flowtron [Official]
Glow-TTS [83] is a flow-based generative model for parallel TTS using a dynamic programming method to achieve the alignment between text and speech. Link: www.github.com/jaywalnut310/glow-tts [Official]
Speech synthesis system libraries
Mozilla TTS is a deep learning library for TTS that includes implementations for Tacotron, Tacotron 2, Glow-TTS and vocoders such as MelGAN, WaveRNN and others. Link: www.github.com/mozilla/TTS [Official]

NeMO is a toolkit that includes solutions for TTS, speech recognition and natural language processing tools as well. Link: www.github.com/NVIDIA/NeMo [Official]
ESPNET-TTS [109] is a toolkit that contains implementations for TTS systems like Tacotron, Transformer TTS, FastSpeech and others. Link: www.github.com/espnet/espnet [Official]
Parakeet is a flexible, efficient and state-of-the-art text-to-speech toolkit for the open-source community. It includes many influential TTS models proposed by Baidu Research and other research groups. Link: www.github.com/PaddlePaddle/Parakeet [Official]
Neural Vocoders
WaveNet [55] is an autoregressive and probabilistic model used to generate raw audio. It can also be conditioned on text to produce the very natural output speech, but its complexity makes it very resource demanding. Link: www.github.com/r9y9/wavenet_vocoder
WaveRNN [88] is a recurrent neural network based vocoder that is able to generate audio faster then real time as a result of its compact architecture. Link: www.github.com/fatchord/WaveRNN
FFTNet [86], inspired by WaveNet also generates the waveform samples sequentially, with the current sample being conditioned on the previous ones, but simplifies its architecture and allows real-time synthesis. Link: www.github.com/syang1993/FFTNet
nv-WaveNet is an open-source implementation of several different single-kernel approaches to the WaveNet variant described by [50]. Link: www.github.com/NVIDIA/nv-wavenet [Official]
LPCNet [89] is a variant of WaveRNN that improves the waveform generation by combining the recurrent neural architecture with linear prediction coefficients. Link: www.github.com/mozilla/LPCNet [Official]
FloWaveNet [94] is a generative model based on flows that can sample audio in real time. Compared to Parallel WaveNet and ClariNet it only requires a training process that is single-staged. Link: www.github.com/ksw0306/FloWaveNet [Official]
Parallel WaveGAN [95] is a vocoder that uses adversarial training and provides fast and lightweight waveform generation. Link: www.github.com/kan-bayashi/ParallelWaveGAN
WaveGlow [95] vocoder borrows from Glow and WaveNet to generate raw audio from Mel spectrograms. It is a flow-based model implemented with a single network. Link: www.github.com/NVIDIA/waveglow [Official]
MelGAN [90] is a GAN-based vocoder that is able to generate coherent waveforms, the model is non-autoregressive and based on convolutional layers. Link: www.github.com/descriptinc/melgan-neurips [Official]
GELP [91] is a parallel neural vocoder utilising generative adversarial networks, and integrating a linear predictive synthesis filter into the model. Link: www.github.com/ljuvela/GELP
SqueezeWave [100] is a lightweight version of WaveGlow that can generate on-device speech output. Link: https://github.com/tianrengao/SqueezeWave [Official]
WaveFlow [96] is a flow-based model that includes WaveNet and WaveGlow as special cases and can synthesise audio faster than real-time. Link: www.github.com/L0SG/WaveFlow
VocGAN [93] is a GAN-based vocoder that can synthesise speech in real time even on a CPU. Link: www.github.com/rishikksh20/VocGAN
WG-WaveNet [97] is composed of a WaveGlow like flow-based model combined with WaveNet based postfilter that can synthesise speech without the need for a GPU. Link: www.github.com/BogiHsu/WG-WaveNet
Speech synthesis challenges
Blizzard Challenge is a yearly challenge in which teams develop TTS systems starting from more or less the same resources, and are jointly evaluated in a large-scale listening test. Link: http://www.festvox.org/blizzard/
Voice Cloning Challenge is a bi-annual challenge in which teams are asked to provide a high-quality solution for cloning the voice of a target speaker within the same language, or cross-lingual. The results are also evaluated in a large scale listening test. Link: http://www.vc-challenge.org/

6. Quality measurements

Although there are no objective measures which can perfectly predict the perceived naturalness of the synthetic output [110, 111], we still need to measure a TTS system's performance. The current approach to doing this is to use *listening tests*. In a listening test, a set of listeners, preferably a large number of native speakers of the target language, are asked to rate the synthetic output in several scenarios using either absolute or relative values. The common setup includes multiple synthesis systems and natural samples. The evaluation can be performed by presenting one or two samples at a time and the listeners rate it by using a Mean Opinion Score (MOS) scale going from 1 to 5, with 5 being the highest value. Or, more commonly used nowadays, in a MUSHRA [112] setup, in which multiple samples are presented the same time and the listeners are asked to order and rate them on a scale of 1 to 100. There is also a preference test setup in which the listeners are asked to choose between two samples according to their preference or adequacy of the rendered speech to the text or speaker identity. The most common evaluation criteria are:

naturalness listeners are asked to rank how close to natural speech is a sample of synthetic output perceived;

intelligibility listeners are asked to transcribe what they hear after playing the sample only once. The transcripts are then compared to the reference transcript and the word error rate is computed;

speaker similarity listeners are presented with a natural sample as reference and a synthetic or natural sample for evaluation. They are asked to rate how similar the identity of the evaluation sample is in comparison to the reference sample.

7. Conclusions and open problems

In this chapter we aimed to provide a high-level indexing of the available methods to generate the voice of an IVA, and to provide the reader with a clear, informed starting point for developing his/her own text-to-speech synthesis system. In the recent years there has been an increasing interest in this domain, especially in the context of vocal chat bots and content access. So that it would be next to impossible to index all the publications and available tools and resources. Yet, we consider that the provided knowledge and minimal scientific description of the TTS domain is sufficient to trigger the interest and application of these methods in the reader's commercial products. It should also be clear that there is still an important trade-off between the quality and the resource requirements of the synthetic voices, and that a very thorough analysis of the applications' specifications and intended use should guide the developer into making the right choice of technology.

We should also point out that, although the recent advancements achieve close to human speech quality, there are still a number of issues that need to be addressed before we can easily say that the topic of speech synthesis has been thoroughly solved. One of these issues is that of *adequate prosody*. When synthesising long paragraphs, or entire books, there is still a lack of variability in the output, and a subset of certain prosodic patterns reemerge. Also, the problem of correctly emphasising certain words, or word groups, such that the desired message is clearly and correctly transmitted is still an open issue for TTS. There is also the problem of mimicking spontaneous speech, where repetitions, elisions, filled pauses, breaks and so on convey the mental process and effort of developing the message and generating it as a spoken discourse.

In terms of speaker identity, the fast adaptation, and also cross-lingual adaptation are of great interest to the TTS community at this point. Being able to copy a person's speech characteristics using as little examples as possible is a daunting task, yet giant leaps have been taken with the NN-based learning. More so, transferring the identity of a person speaking in a language, to the identity of a synthesis system generating a different language is also open for solutions.

On the more far-fetched goals is that of *affective rendering*. If we were to interact with a complete synthetic persona, we would like it to be adaptable to our state of mind, and render compassionate and emphatic emotions in its discourse. Yet the automatic detection and generation of emotions is far from being solved.

Author details

Adriana Stan[1*†] and Beáta Lőrincz[1,2†]

1 Technical University of Cluj-Napoca, Cluj-Napoca, Romania

2 "Babeş-Bolyai" University, Cluj-Napoca, Romania

*Address all correspondence to: adriana.stan@com.utcluj.ro

† These authors contributed equally.

References

[1] J. Benesty, M. M. Sondhi, and Y. A. Huang, *Springer Handbook of Speech Processing*. Secaucus, NJ, USA: Springer-Verlag New York, Inc., 2007.

[2] P. Taylor, *Text-to-Speech Synthesis*. Cambridge University Press, 2009.

[3] "Missing fundamental," en.wikipedia.org/wiki/ Missing fundamental, online; accessed 15-December-2020.

[4] S. King, "Speech Zone - Windowing," speech.zone/windowing/, online; accessed 15-December-2020.

[5] "Fourier analysis," en.wikipedia.org/wiki/Fourier analysis, online; accessed 15-December-2020.

[6] "Cepstrum," en.wikipedia.org/wiki/Cepstrum, online; accessed 15-December-2020.

[7] J. Li, Z. Wu, R. Li, P. Zhi, S. Yang, and H. Meng, "Knowledge-Based Linguistic Encoding for End-to-End Mandarin Text-to-Speech Synthesis," in *Proc.* Interspeech *2019*, 2019, pp. 4494–4498.

[8] M. Nutu, B. Lorincz, and A. Stan, "Deep Learning for Automatic Diacritics Restoration in Romanian," in *Proc. of IEEE 15th International Conference on Intelligent Computer Communication and Processing*, 09 2019, pp. 1–5.

[9] H. Zhang, R. Sproat, A. H. Ng, F. Stahlberg, X. Peng, K. Gorman, and B. Roark, "Neural Models of Text Normalization for Speech Applications," Computational Linguistics, vol. 45, no. 2, pp. 293–337, 2019.

[10] B. Bohnet, R. McDonald, G. Simões, D. Andor, E. Pitler, and J. Maynez, "Morphosyntactic tagging with a meta-BiLSTM model over context sensitive token encodings," in *Proceedings of the 56th Annual Meeting of the Association for Computational Linguistics (Volume 1: Long Papers)*. Melbourne, Australia: Association for Computational Linguistics, Jul. 2018, pp. 2642–2652.

[11] A. Cutler, *Lexical Stress*. John Wiley & Sons, Ltd, 2005, ch. 11, pp. 264–289.

[12] S. Thomas, M. N. Rao, H. A. Murthy, and C. S. Ramalingam, "Natural sounding TTS based on syllable-like units," in *14th European Signal Processing Conference, EUSIPCO 2006, Florence, Italy, September 4–8*, 2006. IEEE, 2006, pp. 1–5.

[13] A. Sokolov, T. Rohlin, and A. Rastrow, "Neural Machine Translation for Multilingual Grapheme-to-Phoneme Conversion," in *Proc.* Interspeech *2019*, 2019, pp. 2065–2069.

[14] "W3C - Speech Synthesis Markup Language (SSML) Version 1.1," https://www.w3.org/TR/speech-synthesis11/, online; accessed 15-December-2020.

[15] T. Mikolov, I. Sutskever, K. Chen, G. S. Corrado, and J. Dean, "Distributed representations of words and phrases and their compositionality," in *Advances in neural information processing systems*, 2013, pp. 3111–3119.

[16] J. Pennington, R. Socher, and C. D. Manning, "Glove: Global vectors for word representation," in *Proceedings of the 2014 conference on empirical methods in natural language processing (EMNLP)*, 2014, pp. 1532–1543.

[17] M. E. Peters, M. Neumann, M. Iyyer, M. Gardner, C. Clark, K. Lee, and L. Zettlemoyer, "Deep contextualized word representations," *arXiv preprint arXiv:1802.05365*, 2018.

[18] J. Devlin, M.-W. Chang, K. Lee, and K. Toutanova, "BERT: Pre-training of

deep bidirectional transformers for language understanding," *arXiv preprint arXiv:1810.04805*, 2018.

[19] X. Huang, A. Acero, H.-W. Hon, and R. Reddy, *Spoken Language Processing: A Guide to Theory, Algorithm, and System Development*, 1st ed. USA: Prentice Hall PTR, 2001.

[20] D. H. Klatt, "Software for a cascade/parallel formant synthesizer," *Journal of The Acoustical Society of America*, vol. 67, 1980.

[21] J. Allen, S. Hunnicut, and D. Klatt, *From Text to Speech: the MITalk System*. Cambridge University Press, 1987.

[22] C. Bickley, K. Stevens, and D. Williams, "A framework for synthesis of segments based on pseudoarticulatory parameters," pp. 211–220, 1997.

[23] K. Richmond, Z.-H. Ling, and J. Yamagishi, "The use of articulatory movement data in speech synthesis applications: An overview - application of articulatory movements using machine learning algorithms [invited review]," Acoustical Science and Technology, vol. 36, no. 6, pp. 467–477, 2015.

[24] D. Hill, "gnuspeech," www.gnu.org/software/gnuspeech/, online; accessed 15-December-2020.

[25] A. Black, P. Taylor, and R. Caley,*The Festival Speech Synthesis System*, University of Edinburgh, 1999.

[26] T. Lambert and A. P. Breen, "A database design for a TTS synthesis system using lexical diphones," in *Proceedings of Interspeech*, 2004.

[27] T. Saito, Y. Hashimoto, and M. Sakamoto, "High-quality speech synthesis using context-dependent syllabic units," in *Proceedings of the Acoustics, Speech, and Signal Processing, 1996. on Conference Proceedings., 1996 IEEE International Conference – Volume 01*, ser. ICASSP '96, 1996, pp. 381–384.

[28] J. Matoušek, Z. Hanzlíček, and D. Tihelka, "Hybrid syllable/triphone speech synthesis," in *Proceedings of the 9th European Conference on Speech Communication and Technology*, Lisbon, Portugal, 2005, pp. 2529–2532.

[29] O. Buza, "Contribut, ii la analizas, şi sinteza vorbirii din text pentru limba română," Ph.D. dissertation, Technical University of Cluj-Napoca, 2010.

[30] R. Stetson, *Motor Phonetics: A Study of Speech Movements in Action*. Oberlin College, 1951.

[31] A. Black and N. Campbell, "Optimising selection of units from speech database for concatenative synthesis," in *Proc. EUROSPEECH-95*, Sep. 1995, pp. 581–584.

[32] A. Hunt and A. Black, "Unit selection in a concatenative speech synthesis system using a large speech database," in *Proc. of ICASSP*, May 1996, pp. 373–376.

[33] Y. Qian, F. K. Soong, and Z. Yan, "A unified trajectory tiling approach to high quality speech rendering," IEEE Transactions on Audio, Speech, and Language Processing, vol. 21, no. 2, pp. 280–290, 2013.

[34] A. Falaschi, M. Giustiniani, and M. Verola, "A hidden Markov model approach to speech synthesis," in *Proceedings of Eurospeech*, vol. 1989, 1989, pp. 2187–2190.

[35] S. King, "An introduction to statistical parametric speech synthesis," Sadhana, vol. 36, p. 837–852, 2011.

[36] H. Zen, T. Toda, M. Nakamura, and K. Tokuda, "Details of Nitech HMM-based speech synthesis system for the Blizzard Challenge 2005," IEICE Trans.

Inf. & Syst., vol. E90-D, no. 1, pp. 325–333, Jan. 2007.

[37] J. Yamagishi, B. Usabaev, S. King, O. Watts, J. Dines, J. Tian, R. Hu, Y. Guan, K. Oura, K. Tokuda, R. Karhila, and M. Kurimo, "Thousands of voices for HMM-based speech synthesis – analysis and application of TTS systems built on various ASR corpora," IEEE Transactions on Audio, Speech and Language Processing, vol. 18, no. 5, pp. 984–1004, July 2010.

[38] J. Lorenzo-Trueba, R. Barra-Chicote, R. San-Segundo, J. Ferreiros, J. Yamagishi, and J. M. Montero, "Emotion transplantation through adaptation in hmm-based speech synthesis," Computer Speech and Language, vol. 34, no. 1, pp. 292–307, 2015.

[39] H. Kawahara, I. Masuda-Katsuse, and A. Cheveigné, "Restructuring speech representations using a pitch-adaptive time-frequency smoothing and an instantaneous-frequency-based F0 extraction: possible role of a repetitive structure in sounds," Speech Communication, vol. 27, pp. 187–207, 1999.

[40] M. Morise, F. Yokomori, and K. Ozawa, "WORLD: A Vocoder-Based High-Quality Speech Synthesis System for Real-Time Applications," IEICE Transactions, vol. 99-D, pp. 1877–1884, 2016.

[41] Q. Hu, K. Richmond, J. Yamagishi, and J. Latorre, "An experimental comparison of multiple vocoder types," in *8th ISCA Workshop on Speech Synthesis*, Barcelona, Spain, August 2013, pp. 155–160.

[42] W. S. McCulloch and W. Pitts, "A logical calculus of the ideas immanent in nervous activity," Neurocomputing: Foundations of Research, p. 15–27, 1988.

[43] M. G. Rahim and C. C. Goodyear, "Articulatory synthesis with the aid of a neural net," in *International Conference on Acoustics, Speech, and Signal Processing*, 1989, pp. 227–230 vol.1.

[44] G. E. Hinton, "Learning multiple layers of representation," Trends in Cognitive Sciences, vol. 11, no. 10, pp. 428–434, 2007.

[45] G. E. Hinton, S. Osindero, and Y.-W. Teh, "A fast learning algorithm for deep belief nets," Neural Computation, vol. 18, no. 7, pp. 1527–1554, 2006.

[46] Z. Ling, L. Deng, and D. Yu, "Modeling Spectral Envelopes Using Restricted Boltzmann Machines and Deep Belief Networks for Statistical Parametric Speech Synthesis," IEEE Transactions on Audio, Speech, and Language Processing, vol. 21, no. 10, pp. 2129–2139, 2013.

[47] H. Zen, A. Senior, and M. Schuster, "Statistical parametric speech synthesis using deep neural networks," in *2013 IEEE International Conference on Acoustics, Speech and Signal Processing*, 2013, pp. 7962–7966.

[48] Z. Wu, O. Watts, and S. King, "Merlin: An open source neural network speech synthesis system." In *Speech Synthesis Workshop*, 2016, pp. 202–207.

[49] O. Watts, G. Henter, J. Fong, and C. Valentini-Botinhao, "Where do the improvements come from in sequence-to-sequence neural TTS?" in *Proc of the 10th ISCA Speech Synthesis Workshop*. International Speech Communication Association, Sep. 2019, pp. 217–222.

[50] S. O. Arik, M. Chrzanowski, A. Coates, G. Diamos, A. Gibiansky, Y. Kang, X. Li, J. Miller, A. Ng, J. Raiman, S. Sengupta, and M. Shoeybi, "Deep voice: Real-time neural text-to-speech," *arXiv preprint arXiv:1702.07825*, 2017.

[51] W. Wang, S. Xu, and B. Xu, "First Step Towards End-to-End Parametric TTS Synthesis: Generating Spectral

Parameters with Neural Attention," in *Interspeech 2016*, 2016, pp. 2243–2247.

[52] J. Sotelo, S. Mehri, K. Kumar, J. F. Santos, K. Kastner, A. C. Courville, and Y. Bengio, "Char2wav: End-to-end speech synthesis," in *5th International Conference on Learning Representations, ICLR 2017, Toulon, France, April 24–26, 2017, Workshop Track Proceedings*. OpenReview.net, 2017.

[53] Y. Wang, R. Skerry-Ryan, D. Stanton, Y. Wu, R. J. Weiss, N. Jaitly, Z. Yang, Y. Xiao, Z. Chen, S. Bengio, Q. Le, Y. Agiomyrgiannakis, R. Clark, and R. A. Saurous, "Tacotron: Towards end-to-end speech synthesis," in *Proc. of Interspeech*, 2017.

[54] D. Griffin and J. Lim, "Signal estimation from modified short-time Fourier transform," IEEE Transactions on Acoustics, Speech, and Signal Processing, vol. 32, no. 2, pp. 236–243, 1984.

[55] A. v. d. Oord, S. Dieleman, H. Zen, K. Simonyan, O. Vinyals, A. Graves, N. Kalchbrenner, A. Senior, and K. Kavukcuoglu, "Wavenet: A generative model for raw audio," *arXiv preprint arXiv:1609.03499*, 2016.

[56] R. J. Weiss, R. Skerry-Ryan, E. Battenberg, S. Mariooryad, and D. P. Kingma, "Wave-tacotron: Spectrogram-free end-to-end text-to-speech synthesis," *arXiv preprint arXiv: 2011.03568*, 2020.

[57] A. Peiró-Lilja and M. Farrús, "Naturalness Enhancement with Linguistic Information in End-to-End TTS Using Unsupervised Parallel Encoding," in *Proc.* Interspeech *2020*, 2020, pp. 3994–3998.

[58] J. Taylor and K. Richmond, "Enhancing Sequence-to-Sequence Text-to-Speech with Morphology," in *Proc.* Interspeech *2020*, 2020, pp. 1738–1742.

[59] J. Shen, R. Pang, R. J. Weiss, M. Schuster, N. Jaitly, Z. Yang, Z. Chen, Y. Zhang, Y. Wang, R. Skerrv-Ryan, R. A. Saurous, Y. Agiomvrgiannakis, and Y. Wu, "Natural TTS Synthesis by Conditioning Wavenet on MEL Spectrogram Predictions," in *2018 IEEE International Conference on Acoustics,* Speech and Signal Processing (ICASSP), 2018, pp. 4779–4783.

[60] H. Tachibana, K. Uenoyama, and S. Aihara, "Efficiently trainable text-to-speech system based on deep convolutional networks with guided attention," in *2018 IEEE International Conference on Acoustics, Speech and Signal Processing (ICASSP)*, 2018, pp. 4784–4788.

[61] W. Ping, K. Peng, A. Gibiansky, S. O. Arik, A. Kannan, S. Narang, J. Raiman, and J. Miller, "Deep voice 3: 2000-speaker neural text-to-speech," Proc. ICLR, pp. 214–217, 2018.

[62] W. Ping, K. Peng, and J. Chen, "Clarinet: Parallel wave generation in end-to-end text-to-speech," *arXiv preprint arXiv:1807.07281*, 2018.

[63] K. Peng, W. Ping, Z. Song, and K. Zhao, "Non-autoregressive neural text-to-speech," in *Proceedings of the 37th International Conference on Machine Learning*, ser. Proceedings of Machine Learning Research, H. D. III and A. Singh, Eds., vol. 119. Virtual: PMLR, 13–18 Jul 2020, pp. 7586–7598.

[64] C. Yu, H. Lu, N. Hu, M. Yu, C. Weng, K. Xu, P. Liu, D. Tuo, S. Kang, G. Lei, D. Su, and D. Yu, "DurIAN: Duration Informed Attention Network for Speech Synthesis," in *Proc.* Interspeech *2020*, 2020, pp. 2027–2031.

[65] J. Shen, Y. Jia, M. Chrzanowski, Y. Zhang, I. Elias, H. Zen, and Y. Wu, "Non-attentive tacotron: Robust and controllable neural tts synthesis including unsupervised duration modeling," 2020.

[66] A. Vaswani, N. Shazeer, N. Parmar, J. Uszkoreit, L. Jones, A. N. Gomez, Ł. Kaiser, and I. Polosukhin, "Attention is all you need," Advances in neural information processing systems, vol. 30, pp. 5998–6008, 2017.

[67] N. Li, S. Liu, Y. Liu, S. Zhao, and M. Liu, "Neural speech synthesis with transformer network," in *Proceedings of the AAAI Conference on Artificial Intelligence*, vol. 33, 2019, pp. 6706–6713.

[68] Y. Ren, Y. Ruan, X. Tan, T. Qin, S. Zhao, Z. Zhao, and T.-Y. Liu, "Fastspeech: Fast, robust and controllable text to speech," in *Advances in Neural Information Processing Systems*, 2019, pp. 3171–3180.

[69] Y. Ren, C. Hu, T. Qin, S. Zhao, Z. Zhao, and T.-Y. Liu, "Fastspeech 2: Fast and high-quality end-to-end text-to-speech," *arXiv preprint arXiv: 2006.04558*, 2020.

[70] Z. Zeng, J. Wang, N. Cheng, T. Xia, and J. Xiao, "Aligntts: Efficient feed-forward text-to-speech system without explicit alignment," in *ICASSP 2020–2020 IEEE International Conference on Acoustics, Speech and Signal Processing (ICASSP)*. IEEE, 2020, pp. 6714–6718.

[71] D. Lim, W. Jang, G. O, H. Park, B. Kim, and J. Yoon, "JDI-T: Jointly Trained Duration Informed Transformer for Text-To-Speech without Explicit Alignment," in *Proc.* Interspeech *2020*, 2020, pp. 4004–4008.

[72] M. Chen, X. Tan, Y. Ren, J. Xu, H. Sun, S. Zhao, and T. Qin, "Multispeech: Multi-speaker text to speech with transformer," *arXiv preprint arXiv: 2006.04664*, 2020.

[73] H. R. Ihm, J. Y. Lee, B. J. Choi, S. J. Cheon, and N. S. Kim, "Reformer-TTS: Neural Speech Synthesis with Reformer Network," Proc. Interspeech *2020*, pp. 2012–2016, 2020.

[74] W.-N. Hsu, Y. Zhang, and J. Glass, "Learning latent representations for speech generation and transformation," *arXiv preprint arXiv:1704.04222*, 2017.

[75] Y.-J. Zhang, S. Pan, L. He, and Z.-H. Ling, "Learning latent representations for style control and transfer in end-to-end speech synthesis," in *ICASSP 2019–2019 IEEE International Conference on Acoustics, Speech and Signal Processing (ICASSP)*. IEEE, 2019, pp. 6945–6949.

[76] G. Sun, Y. Zhang, R. J. Weiss, Y. Cao, H. Zen, A. Rosenberg, B. Ramabhadran, and Y. Wu, "Generating diverse and natural text-to-speech samples using a quantized fine-grained VAE and autoregressive prosody prior," in *IEEE International Conference on Acoustics, Speech and Signal Processing (ICASSP)*. IEEE, 2020, pp. 6699–6703.

[77] Y. Yasuda, X. Wang, and J. Yamagishi, "End-to-End Text-to-Speech using Latent Duration based on VQ-VAE," *arXiv preprint arXiv: 2010.09602*, 2020.

[78] V. Aggarwal, M. Cotescu, N. Prateek, J. Lorenzo-Trueba, and R. Barra-Chicote, "Using VAEs and Normalizing Flows for One-Shot Text-To-Speech Synthesis of Expressive Speech," in *ICASSP 2020-2020 IEEE International Conference on Acoustics, Speech and Signal Processing (ICASSP)*. IEEE, 2020, pp. 6179–6183.

[79] M. Bińkowski, J. Donahue, S. Dieleman, A. Clark, E. Elsen, N. Casagrande, L. C. Cobo, and K. Simonyan, "High fidelity speech synthesis with adversarial networks," *arXiv preprint arXiv:1909.11646*, 2019.

[80] H. Guo, F. K. Soong, L. He, and L. Xie, "A new GAN-based end-to-end TTS training algorithm," *arXiv preprint arXiv:1904.04775*, 2019.

[81] D. J. Rezende and S. Mohamed, "Variational inference with normalizing

flows," *arXiv preprint arXiv:1505.05770*, 2015.

[82] R. Valle, K. Shih, R. Prenger, and B. Catanzaro, "Flowtron: an autoregressive flow-based generative network for text-to-speech synthesis," *arXiv preprint arXiv:2005.05957*, 2020.

[83] J. Kim, S. Kim, J. Kong, and S. Yoon, "Glow-TTS: A Generative Flow for Text-to-Speech via Monotonic Alignment Search," *arXiv preprint arXiv:2005.11129*, 2020.

[84] C. Miao, S. Liang, M. Chen, J. Ma, S. Wang, and J. Xiao, "Flow-TTS: A Non-Autoregressive Network for Text to Speech Based on Flow," in *2020 IEEE International Conference on Acoustics, Speech and Signal Processing (ICASSP)*. IEEE, 2020, pp. 7209–7213.

[85] P. Govalkar, J. Fischer, F. Zalkow, and C. Dittmar, "A Comparison of Recent Neural Vocoders for Speech Signal Reconstruction," in *Proc. 10th ISCA Speech Synthesis Workshop*, 2019, pp. 7–12.

[86] Z. Jin, A. Finkelstein, G. J. Mysore, and J. Lu, "FFTNet: A Real-Time Speaker-Dependent Neural Vocoder," in *IEEE International Conference on Acoustics, Speech and Signal Processing (ICASSP)*, 2018, pp. 2251–2255.

[87] A. Oord, Y. Li, I. Babuschkin, K. Simonyan, O. Vinyals, K. Kavukcuoglu, G. Driessche, E. Lockhart, L. Cobo, F. Stimberg et al., "Parallel WaveNet: Fast high-fidelity speech synthesis," in *International conference on machine learning*. PMLR, 2018, pp. 3918–3926.

[88] N. Kalchbrenner, E. Elsen, K. Simonyan, S. Noury, N. Casagrande, E. Lockhart, F. Stimberg, A. v. d. Oord, S. Dieleman, and K. Kavukcuoglu, "Efficient neural audio synthesis," *arXiv preprint arXiv:1802.08435*, 2018.

[89] J.-M. Valin and J. Skoglund, "LPCNet: Improving neural speech synthesis through linear prediction," in *IEEE International Conference on Acoustics, Speech and Signal Processing (ICASSP)*. IEEE, 2019, pp. 5891–5895.

[90] K. Kumar, R. Kumar, T. de Boissiere, L. Gestin, W. Z. Teoh, J. Sotelo, A. de Brébisson, Y. Bengio, and A. C. Courville, "MelGAN: Generative adversarial networks for conditional waveform synthesis," in *Advances in Neural Information Processing Systems*, 2019, pp. 14 910–14 921.

[91] L. Juvela, B. Bollepalli, J. Yamagishi, and P. Alku, "GELP: GAN-excited linear prediction for speech synthesis from mel-spectrogram," *arXiv preprint arXiv: 1904.03976*, 2019.

[92] R. Yamamoto, E. Song, and J.-M. Kim, "Parallel WaveGAN: A fast waveform generation model based on generative adversarial networks with multi-resolution spectrogram," in *IEEE International Conference on Acoustics, Speech and Signal Processing (ICASSP)*. IEEE, 2020, pp. 6199–6203.

[93] J. Yang, J. Lee, Y. Kim, H.-Y. Cho, and I. Kim, "VocGAN: A High-Fidelity Real-Time Vocoder with a Hierarchically-Nested Adversarial Network," in *Proc.* Interspeech *2020*, 2020, pp. 200–204.

[94] S. Kim, S.-g. Lee, J. Song, J. Kim, and S. Yoon, "FloWaveNet: A generative flow for raw audio," *arXiv preprint arXiv:1811.02155*, 2018.

[95] R. Prenger, R. Valle, and B. Catanzaro, "WaveGlow: A flow-based generative network for speech synthesis," in *IEEE International Conference on Acoustics, Speech and Signal Processing (ICASSP)*. IEEE, 2019, pp. 3617–3621.

[96] W. Ping, K. Peng, K. Zhao, and Z. Song, "WaveFlow: A compact flow-based model for raw audio," in *International Conference on Machine Learning*. PMLR, 2020, pp. 7706–7716.

[97] P. chun Hsu and H. yi Lee, "WG-WaveNet: Real-Time High-Fidelity Speech Synthesis Without GPU," in *Proc.* Interspeech, 2020, pp. 210–214.

[98] W. Song, G. Xu, Z. Zhang, C. Zhang, X. He, and B. Zhou, "Efficient WaveGlow: An Improved WaveGlow Vocoder with Enhanced Speed," in *Proc.* Interspeech, 2020, pp. 225–229.

[99] Z. Zeng, J. Wang, N. Cheng, and J. Xiao, "MelGlow: Efficient Waveform Generative Network Based on Location-Variable Convolution," *arXiv preprint arXiv:2012.01684*, 2020.

[100] B. Zhai, T. Gao, F. Xue, D. Rothchild, B. Wu, J. E. Gonzalez, and K. Keutzer, "SqueezeWave: Extremely Lightweight Vocoders for On-device Speech Synthesis," *arXiv preprint arXiv: 2001.05685*, 2020.

[101] M. Gavrilidou, P. Labropoulou, E. Desipri, S. Piperidis, H. Papageorgiou, M. Monachini, F. Frontini, T. Declerck, G. Francopoulo, V. Arranz, and V. Mapelli, "The META-SHARE Metadata Schema for the Description of Language Resources." in *LREC*, 2012, pp. 1090–1097.

[102] H. Zen, V. Dang, R. Clark, Y. Zhang, R. J. Weiss, Y. Jia, Z. Chen, and Y. Wu, "LibriTTS: A corpus derived from LibriSpeech for text-to-speech," *arXiv preprint arXiv:1904.02882*, 2019.

[103] A. W. Black, "CMU Wilderness Multilingual Speech Dataset," in *IEEE International Conference on Acoustics, Speech and Signal Processing (ICASSP)*, 2019, pp. 5971–5975.

[104] A. Stan, "Recoapy: Data recording, pre-processing and phonetic transcription for end-to-end speech-based applications," *arXiv preprint arXiv:2009.05493*, 2020.

[105] M. Schröder, M. Charfuelan, S. Pammi, and I. Steiner, "Open source voice creation toolkit for the MARY TTS Platform," in *Proc. of Interspeech*, 2011.

[106] B. Potard, M. P. Aylett, D. A. Baude, and P. Motlicek, "Idlak Tangle: An Open Source Kaldi Based Parametric Speech Synthesiser Based on DNN." in *Proc. of Interspeech*, 2016, pp. 2293–2297.

[107] Y. Taigman, L. Wolf, A. Polyak, and E. Nachmani, "Voiceloop: Voice fitting and synthesis via a phonological loop," *arXiv preprint arXiv:1707.06588*, 2017.

[108] R. Valle, J. Li, R. Prenger, and B. Catanzaro, "Mellotron: Multispeaker expressive voice synthesis by conditioning on rhythm, pitch and global style tokens," in *IEEE International Conference on Acoustics, Speech and Signal Processing (ICASSP)*. IEEE, 2020, pp. 6189–6193.

[109] T. Hayashi, R. Yamamoto, K. Inoue, T. Yoshimura, S. Watanabe, T. Toda, K. Takeda, Y. Zhang, and X. Tan, "ESPnet-TTS: Unified, reproducible, and integratable open source end-to-end text-to-speech toolkit," in *IEEE International Conference on Acoustics, Speech and Signal Processing (ICASSP)*. IEEE, 2020, pp. 7654–7658.

[110] Y. Choi, Y. Jung, and H. Kim, "Deep MOS Predictor for Synthetic Speech Using Cluster-Based Modeling," in *Proc.* Interspeech *2020*, 2020, pp. 1743–1747.

[111] M. Wester, C. Valentini-Botinhao, and G. E. Henter, "Are we using enough listeners? No! An empirically-supported critique of Interspeech 2014 TTS evaluations," in *Proc. Interspeech*, Dresden, September 2015, pp. 3476–3480.

[112] M. Schoeffler, S. Bartoschek, F.-R. Stöter, M. Roess, S. Westphal, B. Edler, and J. Herre, "webMUSHRA — A Comprehensive Framework for Web-based Listening Tests," *Journal of Open Research Software*, vol. 6, no. 1, p. 8, Feb. 2018, number: 1 Publisher: Ubiquity Press.

Chapter 3

Virtual Assistants and Ethical Implications

Abhishek Kaul

Abstract

Virtual assistants are becoming a part of our daily life, from our homes to our work. Sometime we may not even know that the customer service agent that we were speaking with is a virtual assistant. These assistants continuously collect information from our interactions and learn many things about us. The information they gather over time is enormous. This chapter introduces the concept of ethics, discusses the ethical principles of virtual assistants, (Transparency, Justice & fairness, Non-maleficence, Responsibly and Privacy). Although there is limited regulation governing these virtual assistants, practical guidelines and recommendations are provided for designers and developers to understand the ethical implications when building a virtual assistant. In this chapter, we also discuss technology and learning techniques for virtual assistants and present examples on how to ensure ethical virtual assistant.

Keywords: virtual assistants, artificial intelligence, ethics, deep learning algorithms, natural language processing, natural language understanding, fair AI, transparent AI

1. Introduction

Organizations are rapidly deploying Virtual Assistants aka bot technology [1] for automating communication, customer service, conversational commerce, product recommendation, education support, financial services, medical services, entertainment, social outreach and self-service tasks. They offer 24/7 service and fulfill the need of millennials [2] for real time responses. Virtual assistants enable organizations to reduce costs, increase brand loyalty and better serve customers. However, virtual assistants are built by humans using artificial intelligence (AI) technologies and have wide ranging ethical implications which are important for organizations and consumers to understand.

Many countries have published AI policy guidelines [3, 4]. These guidelines provide a broad level objective for the use of AI – to ensure human-centric, safe, and trustworthy AI. One of the most important aspect in all guidelines is ethics, "AI should be ethical, ensuring adherence to ethical principles and values". Although, AI by its very nature is a form of statistical discrimination (finding patters in data), the discrimination becomes objectional when it places certain privileged groups at a systematic advantage and certain unprivileged groups at a systematic disadvantage. For example, the loan application algorithm gives higher credit scores to older males due to training bias. Objectional discriminations can arise due to multiple reasons like wrongly defining the business objective [5] of machine learning model, using unrepresentative data or data with existing prejudice [6] for training or by selecting wrong attributes or features of the AI model.

Significant work has been done in the area of Ethically aligned design for Autonomous and Intelligent systems by IEEE [7]; and in the area of Facial recognition technologies [8]; but the area of virtual assistants has seen limited guidelines or regulation. California "Bot Bill [9]" provides only limited protection for consumers in terms of bot self-declaration.

In subsequent sections of this chapter, we first define "What is ethics?" and then discuss on ethical principles for virtual assistants. These principles provide key ethical considerations that designers and consumers should understand. Next we discuss different types of virtual assistants deployed today deep, dive into the technology and learning techniques that make them ethical. Further, we analyze the guidelines and legislations that companies and governments have published. In the last section, we look into what is the probable future of super intelligent virtual assistants.

2. Defining ethics

What is ethics?

As per the Oxford dictionary ethics means "the moral principles that govern a person's behaviour or the conducting of an activity".

Ethics is based on well-founded standards of right and wrong that prescribe what humans ought to do, usually in terms of rights, obligations, benefits to society, fairness, or specific virtues [10].

An ethical virtual assistant should be designed with the ethical standards of the society it affects. These standards should extend to virtual assistants and their creators who should design, build and maintain virtual assistants to ensure that their interactions with consumers foster honesty, loyalty, refrain from doing harm, or fraud and provide the right to privacy.

In the subsequent section, we discuss in detail the ethical principles of virtual assistants.

3. Ethical principles for virtual assistants

AI based virtual assistants ability to act intelligently has long been evaluated by the Turing Test [11] and Loebner Prize [12]. The focus is on intelligence of the system to respond to human questions. Looking through the lens of ethical principles, other questions arise, beyond "What can the virtual assistant answer?" For example, "Does the answer promote consumers interests or business interests like recommending the most profitable product which does not best suited".

In a recent paper published by Jobin, A., Ienca, M. & Vayena, E. on the global landscape of AI ethics guidelines [13], they found globally five emerging ethical principles that are deemed important – Transparency, Justice and fairness, Non-maleficence, Responsibly and Privacy. In this section we interpret these principles with a view on virtual assistants and what considerations should designers, developers and consumers understand when developing and interacting with virtual assistants.

3.1 Transparency

AI transparency refers to the explainability [14], interpretability, disclosure [15] of the algorithmic models including their training data, accuracy, performance, bias and other metrics.

When dealing with virtual assistants, transparency [16] often refers to informing the consumers who they are chatting with ie virtual assistant, not actual human, sharing details on what information can the consumer search, and how his data will be used, stored, analyzed for improving experience.

Brands build trust with consumers by being transparent and honest in communication. Virtual assistants are an extension of brands consumer experience. If virtual assistants impersonate a human, it can lead to poor experience and lack of trust with the brand. This can also be harmful when interacting with consumers on sensitive areas like healthcare or banking.

Designer and developers of virtual assistants should be transparent in disclosing information to consumers in terms of what they can search and disclose how their data would be shared and analyzed. When the consumers know what they can search they will be able to ask questions on the topics that virtual assistant has been trained on and get desirable answers. This will create a delightful experience. Further, consumer should have the choice to opt in their interaction data for other purposes like development of the AI models or for advertisements and more. This will help to gain consumer confidence in virtual assistants and increase adoption. Lastly, consumers should also have the option to connect to a real person, request callback or send an email if they are uncomfortable in interacting with a virtual assistant.

3.2 Justice, fairness and equity

Justice means that AI algorithms are fair and do not discriminate against particular groups intentionally or unintentionally [17]. There have been numerous publications on fairness and how to identify, mitigate bias in Algorithms [18–20]. In case of virtual assistants justice, fairness and equity refer primarily to prioritizing the consumer interests and providing impartial recommendations [21] .

AI models on recommendation generally use techniques of collaborative filtering, ie filtering for consumer preferences based on information gathered from many similar consumers. The models constantly learn from consumer feedback ie likes or dislikes and adjust accordingly.

However these models can be biased based on the consumer training data or based on overarching business rules like recommend the most profitable product. For example, will the virtual assistant recommend the meat which is most expensive and near expiry date or the meat which is cheaper and fresh?

Virtual assistants being viewed favorable towards certain recommendations raises the question on fairness especially for consumers. When virtual assistants are used within an organization, then sometimes recommendation may rule driven, which is as per the employee policy.

Designers and developers should regularly test the virtual assistants against the fairness metrics, publish them to consumers and also give consumers the option to provide feedback on recommendation. The more virtual assistant adapts to consumers interest and provides fair recommendation, the more popular the virtual assistant will become with consumers.

3.3 Non-maleficence

This term is used to define consumers safety, security and the commitment that AI model will not cause harm for example, by spamming, hacking, discrimination, violation of privacy or abuse.

In case of virtual assistants, we focus on abuse and sexual harassment for this principle. Abuse refers to both receiving abuse from consumers and giving back abuse to consumers.

Many times, virtual assistants are at the beginning of a consumers journey, and if the responses are not helpful it leads to frustration and abuse from consumers. Although, virtual assistants are AI models and do not have feelings (like humans), as consumers, we should refrain from abusing since it impacts the way we behave in society and transcends similar behavior towards even our fellow humans.

Designers and developers need to design the conversation experience with consideration that virtual assistants will receive abuse. They should design the conversation flow empathically so that the consumers are provided a positive response and transferred to a more helpful channel like voice or email on request [22].

Another consideration is gender stereo-typing ie the gender of virtual assistant. In many cases, virtual assistants have a default female voice or persona. Designers and developers can provide options to consumers to select the virtual assistant persona and alter language, voice, tone of responses specific to chosen persona.

In a related study on sexual harassment of virtual assistants, "#MeToo: How Conversational Systems Respond to Sexual Harassment [23]" points different behaviors in commercial, supervised and unsupervised learning based virtual assistants. The unsupervised learning based assistants have more freedom in learning from user conversation and responding similarly. In these cases language correction models should also be deployed to protect users from chatbot abuse. For example, Microsoft's Tay chatbot was corrupted in less than 24 hours by self-learning through user conversation [24].

3.4 Responsibility and accountability

Responsibility and accountability refer to the AI acting with integrity, clarifying the attribution of responsibility and data ownership. In case of virtual assistants this refers to being transparent, fair, disclosing information on responsibility, legal liability and data ownership to consumers.

There has been much debate on who is ultimately responsible – is it the AI based virtual assistants or the humans who built it. Generally, terms of service agreement which consumers have to agree before using virtual assistants, define the limitations on responsibilities and liabilities in line with regulations.

Data ownership requires special mention here. Questions typically arise on who owns the data when it is captured and generated during conversation with virtual assistant. For example, new data is generated when a virtual assistant interacts with consumers using voice. It will over time develop data related to consumers preferences (preference in music), personality [25] (words and tone of language), family (number of different voices in family or type of requests made ~ nursery rhymes) and more. Sometimes, organization may have built the business model on leveraging this derived data for profit. For example, Virtual assistant derives data on the age of your children and serves you advertisements on children toothbrush.

Designer and developers should be transparent on data ownership and have an opt in feature, if the consumers want to share this new data generated or want to keep it private. If the business model of the virtual assistant is based on offering free services and leverage consumer data for advertisements, then that should also be transparent to the consumer.

3.5 Privacy

Privacy means that your personal information is kept confidential and only shared with consent. Many countries have passed laws and regulations to protect the privacy of their citizens like General Data Protection Regulation [26]. In relation to Virtual assistants, privacy is often referred in relation to data protection and security.

Deeper questions on privacy for Virtual assistants arise from

- who has access to the conversation transcripts?
- are the transcripts being used to profile the consumers?
- are the transcripts being shared with advertisers?
- are the consumer details anonymized before sharing?
- are the transcripts being used for improving the AI model?
- where are the transcripts stored?
- for how much time are the transcripts stored?
- can the consumer delete the transcripts?
- is the communication channel encrypted?

And so on.

Designer and developers should be transparent on privacy policy and publish it online so that consumers can be informed on how their information is stored and protected. This will also help to develop trust in the virtual assistant and consumers will be more willing to share information if they know that they will be served better.

4. Learning techniques of virtual assistants and ethical considerations

In self-service technology, virtual assistants are on the higher maturity curve and are expected to understand and interact with consumers as "humans" to provide information or take action. If we look under the hood of virtual assistants, then we uncover three basic technology building bocks.

1. Channel of communication – Physical device (Amazon Echo, iPhone Siri), Messaging Platform (Slack, Facebook messenger), Website or App. The channel of communication generally includes voice interaction capability if available.

2. Conversational platform – Brain of the virtual assistant which has the rules and AI technology to understand consumer information and context.

3. Backend Database or Automation/APIs – This is the backend system from where information is retrieved or a specific task is executed. For example calling an API to retrieve weather information for location or setting up an alarm.

In this section, focus will be on the conversational platform which has to be designed with ethical consideration. There are many types of technologies deployed for virtual assistants ranging from simple click based predefined options, to pattern matching, natural language understanding and natural language generation. In the section below, different types conversational platforms are discussed with a view on ethical considerations.

4.1 Commercial virtual assistant platforms

Most commercial virtual assistants use pattern matching and natural language understanding AI models. The primary task of the AI model in this case is to classify intent of the question for pre-defined set of answers. The assistants can also understand specific details in the text like country name or time and more. For example if asked "What is the weather in Singapore?" assistant will classify this as the request to find weather information and also extract Singapore country name. This information will be passed to backend API to retrieve the temperature and presented back as the answer. Example of these virtual assistants used by business are IBM Watson Assistant, Microsoft Bot framework, Amazon Lex, Google Dialog flow and more. Learning on these platform is generally supervised and the knowledge corpus is limited to the business use case. Sometimes, extension of these platforms is done where a large document corpus is ingested and most relevant document is brought forward to the user based on search and retrieval techniques.

In these platforms, it is the role of conversation designer and developer to ensure that the virtual assistant adheres with the ethical principles of Transparency, Justice & fairness, Non-maleficence, Responsibly and Privacy. Further, it is a good practice that document corpus is screened before being ingested into these virtual assistants to ensure relevant and proper responses.

4.2 Mass market virtual assistants

Siri, Alexa and Hey Google are examples of mass market, virtual assistants. These virtual assistants are pre-trained from a large language corpus and have the ability to retrieve personal information from calendar, phonebook, music, credit card and more. The organization developing these Virtual assistants publish their terms of service, privacy policy [27] publicly and it is consumers decision to understand and then interact with them.

The ubiquitous nature of these Virtual assistants poses a bigger question to society on how they should respond to different types of talk ranging from Rude talk, Abusive talk, Romantic talk or Suicidal talk. We discuss below two cases in detail, rude and romantic talk.

Rude Talk – the virtual assistants tend to respond back positively with information without prompts for polite or rude requests. This has an influence on manners especially in case of younger consumers [28]. For example "Alexa can you please tell me the weather forecast for today" or "Alexa weather forecast today" – the answer would be the same. These assistants should try to add nicer words like "Thank you" when consumers say "please".

Romantic Talk and Gender – when asked on gender, the virtual assistants tend respond on gender neutrality. However, by default they respond in a female voice. In the article by Jessi Hempel [29] in Wired she explains that people tend to perceive female voices as helping us to solve our problems. This also opens the door to romantic talk [30] for female persona based virtual assistants. Most of assistant are trained to handle this conversations by evading, or positively responding to consumers, but they rarely respond negatively [31]. This does extend in some cases to general acceptance of sexual harassment for assistants.

4.3 Niche virtual assistants [open domain]

A special mention here to Virtual assistants who can talk about anything in the open domain. These assistants are trained using sophisticated deep learning AI models (un-supervised learning), have billions of parameters and are closest to how

DOI: http://dx.doi.org/10.5772/intechopen.95479

a human would sensibly and specifically answer questions. Many gigabytes of training data (dialog response) is ingested in these AI models and it generates the answers (natural language generation) based on learning. Example of these assistants are

- Meena [32] - trained on 341 GB of text, filtered from public domain social media conversations.

- DialoGPT [33] - large pre-trained **dialog** response model trained on 147 M multi-turn **dialogs** extracted from Reddit discussion threads.

- Mitsuku [34] - although this virtual assistant uses pattern matching techniquc, it has won many competitions.

- Cleverbot [35] - this searches through its saved conversations, and responds to the input by finding how a human responded to that input when it was asked, in part or in full

Other than Mitsuku which uses supervised learning, for other virtual assistants, it is difficult predict responses since they are learning from dialog-response corpus. In these cases, it would be beneficial to have a language filter that checks for ethical considerations like abuse words and more before presenting the answers to consumers.

5. Guidelines and legislations

Many countries have published AI policy guidelines. These guidelines provide a broad level objective for the use of AI – to ensure human-centric, safe, and trustworthy AI. Most guidelines make the organization using AI responsible and accountable for their decisions and ask for the same ethical standards in AI-driven decisions as in human-driven decisions.

General key points achieved from global guidelines are:

- it should be lawful, complying with all applicable laws and regulations;

- it should be ethical, ensuring adherence to ethical principles and values; and

- it should be robust, both from a technical and social perspective since, even with good intentions, AI systems can cause unintentional harm.

Specially for virtual assistants, as defined above, five emerging ethical principles that are deemed important – Transparency, Justice & fairness, Non-maleficence, Responsibly and Privacy.

Legislation has been passed in California [36] to ensure Transparency of Virtual assistants. This law makes it mandatory for Virtual assistants (Bots) to disclose that they are not a real person and are virtual. Many other countries are passing laws and issuing guidelines to make it mandatory for Designers and developers to develop ethical Virtual assistants.

Many commercial organization have also issued their ethical guidelines. IBM has established an Ethics Board [37] which provides governance, review and decision making processes for IBM on ethics policies, practices, communication, research, products and services. They have also published open source toolkits which designers and developers can use to test whether there machine learning, AI models are transparent, fair and explainable.

Google Deepmind [38] has established a focus group which focuses on ethical standards and safety. They look it from the lens of Privacy, transparency and fairness; AI morality and values; Governance and accountability; AI & worlds complex challenges, Misuse and unintended consequences; Economic impact and inclusion.

Microsoft [39] has issued guidelines for responsible bots, which are aimed at helping designers and developers to design a bot that builds trust in the company and service that the bot represents.

Many other companies have issued guidelines to ensure that the Virtual assistants developed on their platform maintain high ethical standards [40] like use supervised learning, divert issues on serios topics, do not spam users, keep user privacy, no advertisements and so on.

6. What comes next

Virtual Assistant AI technology is growing at exponential pace. In the next few years we will have virtual assistants that surpass an average humans ability to respond sensibly and specially to a consumers question. Nick Bostrom [41] presents an interesting perspective on super intelligent moral thinking. In the distant future, as AI capabilities surpasses human intelligence, it could do better than human thinkers and have the correct answers on ethics by weighing up evidence. We have already started seeing the initial versions of these intelligent machines.

IBM Debater [42] is an example of super intelligent system. This AI system can independently debate a human and provide persuasive arguments on complex topics. The system is able to listen and understand a long spontaneous speech, model human dilemmas to form an argument and generate a whole speech of an opinion and deliver is persuasively. The system has participated live and won many debate competitions.

Another example is from Soul Machines [43]. It provides Digital people ie animation of life like people on the screen. These screen animations of people is similar to an actual human who speak with expressions (eye, lips and facial movements). This provides a comfort feeling when interreacting with virtual assistant.

As virtual assistants become a part of our daily life, ethical issues regarding virtual assistants will continue to grow. It is important for the society at large to discuss and agree on the ethical principles of Transparency, Justice & fairness, Non-maleficence, Responsibly and Privacy for virtual assistants.

Conflict of interest

The views expressed in this chapter are my own and are not representative of my employer.

Notes/thanks/other declarations

I thank Ali Soofastaei, who has been my mentor and guide for the initiative of publishing this chapter on Virtual Assistants and Ethical Considerations.

Author details

Abhishek Kaul
IBM, Singapore

*Address all correspondence to: abhishekkaul@gmail.com

References

[1] Building better bots with Watson Conversation. Available from: https://www.ibm.com/blogs/watson/2016/07/building-better-bots-watson-conversation/ [Accessed: 2020-12-10]

[2] Why Millennials Have Higher Expectations for Customer Experience Than Older Generations. Available from: https://www.forbes.com/sites/nicolemartin1/2019/03/26/why-millennials-have-higher-expectations-for-customer-experience-than-older-generations/ [Accessed: 2020-12-10]

[3] Ethics guidelines for trustworthy AI. Available from: https://ec.europa.eu/digital-single-market/en/news/ethics-guidelines-trustworthy-ai [Accessed: 2020-12-10]

[4] Guidance Data Ethics Framework. Available from https://www.gov.uk/government/publications/data-ethics-framework [Accessed: 2020-12-10]

[5] The case for fairer algorithms. Available from: https://medium.com/@Ethics_Society/the-case-for-fairer-algorithms-c008a12126f8 [Accessed: 2020-12-10]

[6] InCoding — In The Beginning Was The Coded Gaze. Available from: https://medium.com/mit-media-lab/incoding-in-the-beginning-4e2a5c51a45d [Accessed: 2020-12-10]

[7] Ethics In Action | Ethically Aligned Design. Available from: https://ethicsinaction.ieee.org/ [Accessed: 2020-12-10]

[8] The ethical questions that haunt facial-recognition research. Available from: https://www.nature.com/articles/d41586-020-03187-3 [Accessed: 2020-12-10]

[9] An act to add Chapter 6 (commencing with Section 17940) to Part 3 of Division 7 of the Business and Professions Code, relating to bots. Available from: https://leginfo.legislature.ca.gov/faces/billTextClient.xhtml?bill_id=201720180SB1001 [Accessed: 2020-12-10]

[10] What is Ethics? Available from: https://www.scu.edu/ethics/ethics-resources/ethical-decision-making/what-is-ethics/ [Accessed: 2020-12-10]

[11] What is Ethics? Available from: https://plato.stanford.edu/entries/turing-test/ [Accessed: 2020-12-10]

[12] What is Ethics? Available from: https://en.wikipedia.org/wiki/Loebner_Prize [Accessed: 2020-12-10]

[13] Jobin, A., Ienca, M. & Vayena, E. The global landscape of AI ethics guidelines. Nat Mach Intell 1, 389-399 (2019). https://doi.org/10.1038/s42256-019-0088-2

[14] AI Explainability 360 – Resources. Available from: http://aix360-dev.mybluemix.net/resources#guidance [Accessed: 2020-12-10]

[15] Introduction to AI FactSheets. Available from: https://aifs360.mybluemix.net/introduction [Accessed: 2020-12-10]

[16] 5 Chatbot Code Of Ethics Every Business Should Follow. Available from: https://botcore.ai/blog/5-chatbot-code-of-ethics-every-business-should-follow/ [Accessed: 2020-12-10]

[17] How to Fight Discrimination in AI. Available from: https://hbr.org/2020/08/how-to-fight-discrimination-in-ai [Accessed: 2020-12-10]

[18] AI Fairness 360. Available from: https://aif360.mybluemix.net/ [Accessed: 2020-12-10]

[19] Fairlearn. Available from: https://github.com/fairlearn/fairlearn [Accessed: 2020-12-10]

[20] Fairness Indicators. Available from: https://github.com/tensorflow/fairness-indicators [Accessed: 2020-12-10]

[21] The code of ethics for AI and chatbots that every brand should follow. Available from: https://www.ibm.com/blogs/watson/2017/10/the-code-of-ethics-for-ai-and-chatbots-that-every-brand-should-follow/ [Accessed: 2020-12-10]

[22] Hard questions about bot ethics. Available from: https://medium.com/slack-developer-blog/hard-questions-about-bot-ethics-4f80797e34f0 [Accessed: 2020-12-10]

[23] #MeToo Alexa: How Conversational Systems Respond to Sexual Harassment. Available from: https://www.aclweb.org/anthology/W18-0802/ [Accessed: 2020-12-10]

[24] Twitter taught Microsoft's AI chatbot to be a racist asshole in less than a day. Available from: https://www.theverge.com/2016/3/24/11297050/tay-microsoft-chatbot-racist [Accessed: 2020-12-10]

[25] Introducing Amazon Halo – Measure body composition, activity, sleep, and tone of voice - Winter + Silver – Medium. Available from: https://www.amazon.com/Amazon-Halo-Fitness-And-Health-Band/dp/B07QK955LS [Accessed: 2020-12-10]

[26] General Data Protection Regulation. Available from: https://gdpr-info.eu/ [Accessed: 2020-12-10]

[27] Alexa and Echo devices are designed to protect your privacy. Available from: https://www.amazon.com/b/?node=19149155011 [Accessed: 2020-12-10]

[28] Amazon Echo Is Magical. It's Also Turning My Kid Into an Asshole. Available from: https://hunterwalk.com/2016/04/06/amazon-echo-is-magical-its-also-turning-my-kid-into-an-asshole/ [Accessed: 2020-12-10]

[29] Siri and Cortana Sound Like Ladies Because of Sexism. Available from: https://www.wired.com/2015/10/why-siri-cortana-voice-interfaces-sound-female-sexism/ [Accessed: 2020-12-10]

[30] Designing an Ethical Chatbot. Available from: https://www.infoq.com/presentations/designing-chatbot-ethics/ [Accessed: 2020-12-10]

[31] We tested bots like Siri and Alexa to see who would stand up to sexual harassment. Available from: https://qz.com/911681/we-tested-apples-siri-amazon-echos-alexa-microsofts-cortana-and-googles-google-home-to-see-which-personal-assistant-bots-stand-up-for-themselves-in-the-face-of-sexual-harassment/ [Accessed: 2020-12-10]

[32] Towards a Conversational Agent that Can Chat About…Anything. Available from: https://ai.googleblog.com/2020/01/towards-conversational-agent-that-can.html [Accessed: 2020-12-10]

[33] DialoGPT: Toward Human-Quality Conversational Response Generation via Large-Scale Pretraining. Available from: https://www.microsoft.com/en-us/research/project/large-scale-pretraining-for-response-generation/ [Accessed: 2020-12-10]

[34] Mitsuku. Available from: https://en.wikipedia.org/wiki/Mitsuku [Accessed: 2020-12-10]

[35] Cleverbot. Available from: https://en.wikipedia.org/wiki/Cleverbot [Accessed: 2020-12-10]

[36] An act to add Chapter 6 (commencing with Section 17940) to

Part 3 of Division 7 of the Business and Professions Code, relating to bots. Available from: https://leginfo.legislature.ca.gov/faces/billTextClient.xhtml?bill_id=201720180SB1001 [Accessed: 2020-12-10]

[37] AI Ethics. Available from: https://www.ibm.com/artificial-intelligence/ethics [Accessed: 2020-12-10]

[38] Exploring the real-world impacts of AI. Available from: https://deepmind.com/about/ethics-and-society [Accessed: 2020-12-10]

[39] Responsible bots: 10 guidelines for developers of conversational AI. Available from: https://www.microsoft.com/en-us/research/publication/responsible-bots/ [Accessed: 2020-12-10]

[40] Ethics and Chatbots. Available from: https://medium.com/pandorabots-blog/ethics-and-chatbots-8d4aab75cca [Accessed: 2020-12-10]

[41] The ethics of artificial intelligence. Available from: https://www.nickbostrom.com/ethics/artificial-intelligence.pdf [Accessed: 2020-12-10]

[42] Project Debater. Available from: https://www.research.ibm.com/artificial-intelligence/project-debater/ [Accessed: 2020-12-10]

[43] Soul Machines. Available from: https://www.soulmachines.com/ [Accessed: 2020-12-10]

Chapter 4

Group-Assign: Type Theoretic Framework for Human AI Orchestration

Aik Beng Ng, Simon See, Zhangsheng Lai and Shaowei Lin

Abstract

In today's Information Age, we work under the constant drive to be more productive. Unsurprisingly, we progress towards being an AI-augmented workforce where we are augmented by AI assistants and collaborate with each other (and their AI assistants) at scale. In the context of humans, a human language suffices to describe and orchestrate our intents (and corresponding actions) with others. This, however, is clearly insufficient in the context of humans and machines. To achieve this, communication across a network of different humans and machines is crucial. With this objective, our research scope covers and presents a type theoretic framework and language built upon type theory (a branch of symbolic logic in mathematics), to enable the collaboration within a network of humans and AI assistants. While the idea of human-machine or human-computer collaboration is not new, to the best of our knowledge, we are one of the first to propose the use of type theory to orchestrate and describe human-machine collaboration. In our proposed work, we define a fundamental set of type theoretic rules and abstract functions *Group* and *Assign* to achieve the type theoretic description, composition and orchestration of *intents* and *implementations* for an AI-augmented workforce.

Keywords: AI Augmentation, Human Computation, Human AI Collaboration, Human AI Framework, Artificial Intelligence

1. Introduction

The nature of work is always transformed when automation is introduced. Looking back in history, automation has typically served to reduce or eliminate the need for manual work. From hand-delivered messages to telegraph, written communications in the past has required much manual labour. Today, email and a slew of instant messaging platforms get the same job done instantaneously and better. In recent years, highly advanced forms of automation involving AI are making headways into the mainstream workplace. Examples include manufacturing robots learning how to perform bin picking, robot patrol enforcing social distancing in midst of a virus situation [1] and many more. Automation therefore has an overall effect of moving human workers up the cognitive value chain, shifting towards increasingly managerial and strategic roles that are more knowledge-based. The reason for this is apparent as automation introduces characteristics such as being stronger and more tireless relative to human workers, and thereby allowing businesses to do more. Taking a step further, AI is also beginning to encroach into the cognitive realm at work whereby as an instance, an insurance company reportedly replaced its employees with an AI system [2]. All things considered; it is understandable why the workforce is under a constant drive to do more.

Generally, however, automation is well-suited only for tasks that are repeatable within a fixed context. Contrast to this, the handling of work tasks across shifting contexts is not a good candidate for automation. As an example, in the face of a widespread consumer behaviour change due to a pandemic, AI models supporting sentiment analysis, fraud detection, marketing and inventory management operations no longer behaved as expected. The article [3] writes:

> *What's clear is that the pandemic has revealed how intertwined our lives are with AI, exposing a delicate codependence in which changes to our behavior change how AI works, and changes to how AI works change our behavior. This is also a reminder that human involvement in automated systems remains key.*

Though AI models are designed for robustness to changes in the incoming data, situations like these reveal the AI models' brittleness when there is a significant shift in input data distribution. This is termed as *out-of-distribution* (OOD). This means that the input data at point of inference is no longer the same as the training data's distribution that the AI model learnt from. This has the negative effect where the AI model not only potentially makes a mistake on OOD inputs, but even confidently classifying it as a known class. Clearly, this is undesirable and in critical deployments, the mistake can be costly. Clearly, this is undesirable and in critical deployments, the mistake can be costly.

To counter this, there are research efforts looking into OOD detection. To illustrate with an example, a deep generative model for OOD detection was trained using in-distribution genomic sequences [4], with the log-likelihoods plotted for both in-distribution and OOD inputs. We can see from the results (**Figure 1**) that the histogram of log-likelihood overlaps significantly for both in-distribution and OOD inputs, thus showing the model's inability to differentiate between in-distribution and OOD. The authors further note that their observations are not in isolation and are congruent with earlier works using image data.

Naturally, artificial general intelligence (AGI) comes to mind when we broach the topic of narrow AI (as afore discussed). We can think of AGI as the AI's ability being on par or exceeding that of a human's ability to learn, understand and perform intellectual tasks. To conceptualise the relationship of narrow AI and AGI, one may think of them as two sides of the same coin or rather, both ends of the same AI spectrum. Simply put, advancements in narrow AI will evolve towards AGI ultimately. Consensus indicates that AGI is not here today [5, 6] and predictions for the

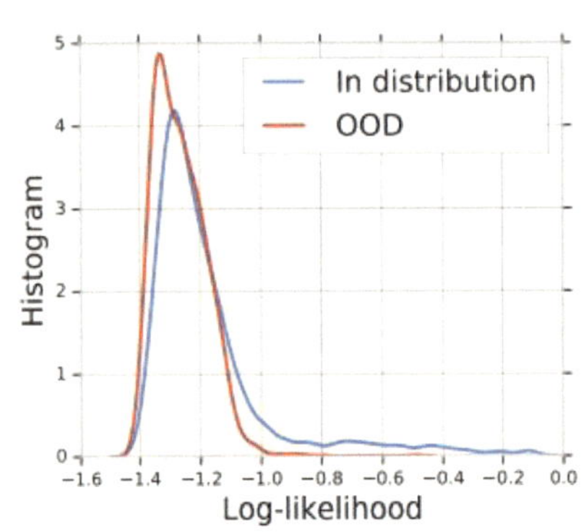

Figure 1.
Log-likelihood hardly separates in-distribution and OOD inputs, adopted from [4].

advent of AGI ranges widely anywhere from a few years to many decades away. Towards this goal, there are research directions on multiple fronts such as to progress deep learning from its current System 1-like abilities to be System 2-capable [7], improving language understanding through increasingly larger and complex language models such as BERT [8] (and its variants) and GPT-3 [9], etc. These are tangible indications that AGI is still some way off.

Meanwhile, we believe there is a complementary and parallel research direction to advancing narrow AI towards AGI. And that is human AI collaboration in which we exploit AI (be it narrow or AGI) to augment our natural abilities. In this sense, for as long as work is envisioned to involve humans, the pursuit for human AI collaboration remains a valuable research direction which will only be more relevant and strengthened further by AI's advancements and increasing pervasiveness at work. In the following sections, we will progressively present our proposed framework designed to enable the collaboration within a network of humans and AI. To begin, we will first discuss some background concepts relevant towards our proposed work.

2. Type theory: formal language of terms and types

Type theory is a branch of symbolic logic in mathematics. The theory of types was conceived to address Russell's paradox arising from naïve set theory, which says that any definable collection is a set. If we define a set of all sets that do not contain themselves, the paradox is illustrated when the set is not a member of itself and therefore needs to contain itself, which then leads to a contradiction of its own definition (Eq. (1)).

$$Let\ S = \{x \in Set | x \notin x\}. Then\ S \in S \Leftrightarrow S \notin S. \tag{1}$$

We can think of type theory as a formal language, complete with a set of rules that inform us on the construction and computations for strings of symbols which we will further introduce hereon the concept of *terms* and *types*.

$$a : A \tag{2}$$

We begin with a basic representation (Eq. (2)), more formally referred to as a judgement, often encountered in type theory, which simply means that a is a term of type A. We can also think of this as a is an element of A. To illustrate for further clarity, we define some examples of terms and types using some real-world objects:

- $red, green, yellow, orange : Colour$
- $mary, john, isaac, caleb : Name$
- $apple, strawberry, banana, orange : Fruit$
- $chicken, fish, beef, turkey : Meat$

In type theory, every term must have a type as seen in List 2. For ease of understanding, we can loosely correspond this to the set theoretic statement $a \in A$ where red, green, yellow and orange are all members of type Colour. Another correspondence can be thought of as *propositions as types* under the Curry-Howard isomorphism [10], where A is a proposition, then a validates the existence of A. To provide the evidence, we will have to perform some mathematical operation to construct the object, that is the term inhabiting the type. Here we observe that *constructivism* is a foundational aspect of type theory, meaning we cannot just assume that objects exist through means such as "Suppose that there exists such an object … ." or by proving the existence of some object through deriving a contradiction from the assumption that object does not exist.

Like other concepts in mathematics, the construction of an object in type theory is governed by rules and we will focus on the introduction of the relevant concepts in the following sections.

2.1 Function type

To start, we look at the *function type*. Given types A and B, we can construct the type $A \to B$ of functions mapping from domain A to codomain B. If we define a function f of type $A \to B$ and apply to the term a of type A, we can obtain a term $f(a)$ of type B (also can be written as $f(a) : B$). For function types, the mapping between the domain to the codomain is constant and fixed. Let us look at a more relatable and practical example of a function type:

- We define a function *head* that returns the first element of a list.
- *head* belongs to the type $List\, a \to a$
- We apply *head* to a list $[1, 2, 3]$ and result is 1.
- If we apply *head* to a list, the result will be.

Hence, a function type will always map from some domain A to the codomain B.

2.2 Dependent product type

Next, we introduce the *dependent product type* for which its terms are functions whose type of codomain varies depending on the term of the domain that the function is applied to. It is also referred to as a *dependent function type* or $\prod$-*type*. Given a type A and a family of types $B : A \to U$, we can construct the type of dependent products $\prod_{x:A} B(x) : U$ where U is known as *universe* whose elements are types. The dependent product type is often used in type theory, and we can think of it as a more generalised form of the function type. The main difference lies in B being a constant family in the function type, such that $\prod_{x:A} B \equiv A \to B$.

To illustrate, let us define f as a dependent function of type $\prod_{x:A} B(x)$ and apply term a of type A. The result is such that we obtain a term $f(a)$ of type $B(a)$ (also can be written as $f(a) : B(a)$). We further provide a more relatable example of a dependent product type as follow:

- We define a dependent function *intOrString* of type $\prod_{x:Boolean} intOrString(x)$ that returns an integer or string depending on the input *true*,*false* : *Boolean*.
- We further define terms and types as 11 : *intOrString*(*true*) and *'hello'* *intOrString* (*false*).
- If we apply *intOrString* to the term *true* of type *Boolean*, an integer 11 is returned.

- If we apply *intOrString* to the term *true* of type *Boolean*, a string *'hello'* is returned.

- Note that *intOrString*(*true*) and *intOrString*(*false*) are different types.

Hence, a dependent product type will map to a different codomain depending on the input term.

2.3 Propositions as types

Earlier on, we briefly talked about the correspondence referred to as *propositions as types*. To validate the truth about a proposition, it means that the corresponding type needs to be inhabited by some term and this is the *evidence* (or *witness*) to the proposition. Generally, the evidence will not be constructed explicitly but rather, translated from proofs into a term of a type and in this sense, it feels like classical set theory reasoning. However, a proposition in type theory goes beyond being true or false, to being a collection of all possible evidence towards the proposition's truth. This mirrors much of our real-world work scenarios, in the sense that there is often more than one correct (true) way of fulfilling a task.

Furthermore, the correspondence between type theoretic and logic operations (**Table 1**) allows us to syntactically construct a type theoretical operation with the semantics of the corresponding logical operation. This is significant because with the ability to correspond between type theoretical and logical operations, the evidence (or proofs) are therefore first-class mathematical objects instead of being just a means for communicating mathematics.

Although it may not be immediately apparent, what we just discussed has impactful implications, mainly:

- Logical operations are integrated within the type theoretical operations, thus combining semantics and syntax. Hence, under the paradigm of propositions as types, a proposition is true and valid when we provide a term to the type. In other words, the type (proposition) is now inhabited by a term (evidence).

- As we are operating with first-class mathematical objects within type theory, this introduces an important aspect: Computability. This gives rise to a further correspondence which is termed as evidence (or proofs) as programs.

- Due to the constructivism nature of type theory, terms are constructed through a set of rules introduced earlier within Section 2. This introduces another important aspect: Explainability.

Logical	Type Theoretic
True	1
False	0
Not A	$A \to 0$
A and B	$A \times B$
A or B	$A + B$
A implies B	$A \to B$
A if and only if B	$(A \to B) \times (B \to A)$

Table 1.
Correspondence of logical and type theoretical operations.

2.4 Reasoning through structured types

Type theory can be viewed as a mathematical formalisation for a programming language. Examples of such programming languages include Agda, Coq, Haskell and more. One notable usage is in proof assistants, that resulted in a verifiable proof of the four colour theorem [11] well over a century after its introduction in 1852. Another notable usage is in formal program verification, which is a software programming paradigm that ensures that the resulting computer program has the rigour of a mathematical proof. This is achieved through specifying how a program should behave and ensuring that it works as specified, which is synonymous with the creation and proof of a mathematical model. Beyond the guarantee of the program's correctness, this has significant implications on cyber security in our highly connected digital society.

Though dependently typed functional programming is not mainstream at the point of writing, it is on the rise and initiatives such as CompCert [12] are active in taking functional programming forward. In concluding Section 2, we find the following quote [13] useful as a succinct summary of type theory:

In type theory, unlike set theory, objects are classified using a primitive notion of type, similar to the data-types used in programming languages. These elaborately structured types can be used to express detailed specifications of the objects classified, giving rise to principles of reasoning about these objects.

3. Group-assign: type theoretic framework for human AI orchestration

Having discussed the background and relevant concepts, we will describe our proposed work hereon. We will first start off with a summary of what the framework is and what it does in Section 3.1. Following this, we will further describe the framework details, methodology and associated terminologies over the subsequent sections. We will also openly discuss about the design considerations that influence the current version of our proposed framework. This is done with the key purpose for sharing our research thoughts through the journey of developing our framework, to better inform future interested parties on how they can leverage and further our proposed work.

3.1 Framework overview and contribution

While the idea of human-machine or human-computer collaboration is not new and different ideas have been proposed. To the best of our knowledge, we are one of the first to propose the use of type theory as a language to orchestrate and describe human-machine collaboration. In our proposed framework, we define a fundamental set of type theoretic rules:

- Base form of intent representation.
- An intent can be applied to different data.
- An intent may result in any number of possible implementations.
- An intent may be composed of one or more constituent intents.

We also define abstract functions for *Group* and *Assign* as base methodologies within the framework to handle data and assigned towards associated implementations.

As an implementation to the type theoretic framework, we develop a prototype using Python that allows us to orchestrate independent declaration of intent(s) and instantiating the *intent(s)* with associated data and implementations, visualised as a simple directed graph that can be recursively built upon a *intent-data-implementation* pattern. Collectively, this graph represents a work plan (e.g. Running a fast food restaurant) in the real world. Each node symbolises some real-world human intent, data group, implementation. For example, "Cook Burger Patty" is an intent that can be instantiated with "Chicken", "Beef" as data groups associated to "Ten steps to cook a burger patty" as an implementation.

3.2 It all begins with an intent

In the context of our proposed work, we define intent simply as "the desire to do something (carry out an implementation)". It is beyond the scope, however, to discuss or quantify intent from a philosophical or psychological view. Before we do something, we first have the intent to do so and the intent does not always necessarily lead to any tangible implementation. Here, we introduce the distinction between intent and implementation.

Work tasks often involve multiple actions and, in this sense, can be considered complex. In undertaking the task, we form an overall intent (which is to complete the task) comprising of constituent intents, which together represents an abstract plan to manage the task. For example, to set up a meeting, we will need to check for the meeting room availability, attendees' availability and then determine the best common time slot. The overall intent in this example is "set up a meeting" with the rest being constituent intents. While "check for meeting room availability" and "check for attendees' availability" can be independent, "determine best common time slot" will depend on these two constituent intents. A constituent intent may depend on one or more other constituent intents or it may also be independent from (existing alongside) other constituent intents. Here, we introduce the notion of a hierarchy of constituent intents within the context of an overall intent.

It can be challenging when we talk about intents. Horizontally across a company, different people in the similar job tiers can have different views about the same thing. Vertically, people across the job tiers will see things at different granularity. Using the same illustration of setting up a meeting, a manager may just instruct the team to set up a meeting. The team member in charge will probably add more constituent intents such as checking for meeting room and attendees' availability because "set up a meeting" is insufficient to fulfil the task. This is an example of vertical granularity differences. Given if another team member is put in charge, he/she may also handle it differently and perhaps add "Cater for coffee and tea" as a constituent intent. This is an example of horizontal diversity. Therefore, to achieve the overall intent (some collective goal), it is important to have the ability to connect diverse and distributed intents in a robust manner.

With these design considerations in mind, framework design principles are summarised as follow:

- Intent and implementation are distinct.
- An intent (within a context) can contain a hierarchy of constituent intents.
- An intent may depend on other constituent intents.
- An intent may have one or more possible implementations.
- Intents should be connected in a robust manner, enabling explanability.

3.3 What is the language for connecting intents?

We earlier discussed about the importance of connecting intents. And by connecting intents, we are composing some work plan. More generally, we are composing a structure and examples abound as we live in a world filled with structures. Examples of structures exist in buildings, deoxyribonucleic acid (DNA), literature, music and many more. The principle of compositionality [14] states that:

> *For every complex expression e in language L, the meaning of e in L is determined by the structure of e in L and the meanings of the constituents of e in L.*

Reasoning is not monolithic and whether as an individual or a team, reasoning is compositional in nature. As we saw earlier, works in AI (both symbolic and neural) are also looking to emulate this behaviour within their AI models. However, intents are intangible and formless. We cannot know what another's intent is unless it is expressed. From a human to human perspective, we compose expressions using some language (e.g. English, Chinese, French, German, etc.) to convey our intents to each another, and the success of it depends on both parties understanding the language as well as whether the expression is well-formed. This takes place so commonly in our daily lives that most of us likely have taken for granted the underlying significance. Hence, an expression is a proxy of our intent and the language is what enables the connection of intents.

Progressing into a future where humans and AI collaborate, will a human language suffice? The answer is clearly no. This is where we believe type theory will serve a suitable and important role in our proposed framework as the language (syntax) that allows users of the framework to define and connect intents and associated implementations (semantics) in a principled way.

3.4 Intents as types can be understood by machines

Humans are not precise and often ambiguous in expressing our intents. Clearly, there is no metric for compatibility and level of abstraction when it comes to human intents. The level and the type of details we deem important and sufficient vary accordingly based on our experience. But with machines, precision and non-ambiguity is critical for things to work.

We believe type theory serves a key role in our proposed framework as the language (syntax) that bridges humans and machines. By embedding human-expressed intents within our type theoretic framework, we posit that the expressiveness (and ambiguity) of humans can be preserved while simultaneously having the precision that machines require in order to function.

This allows users of our framework to define and connect human intents and associated machine implementations (semantics) in a principled and precise manner that also allows for diversity and distributed contributions from multiple parties as is reflective of real world conditions.

Earlier, we presented the idea of correspondence in type theory such as "propositions as types" and "proofs as programs". In our proposed framework, we further introduce a correspondence termed *intents as types* (**Figure 2**).

$$implementation : Intent \tag{3}$$

Recall we established that an intent is distinct from its implementation and an intent may have one or more implementations. This structure is a natural correspondence to the basic representation of term and type (Eq. (2)) earlier introduced

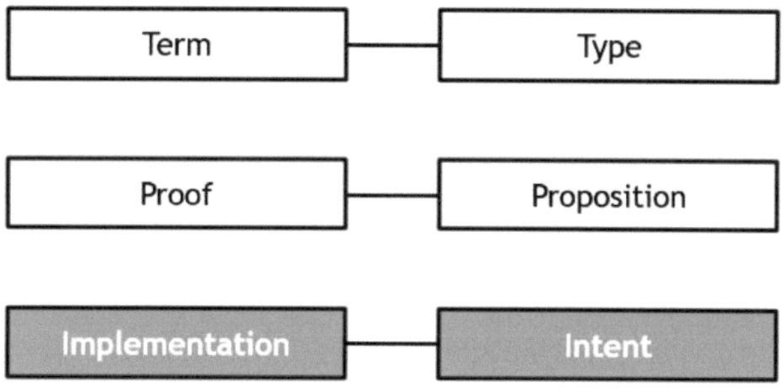

Figure 2.
Intents as types.

and we will represent the base form of an intent in a similar type theoretic manner (Eq. (3)). Hence, we can easily understand this correspondence as:

- Intents as types
- Implementations as terms

By establishing intents as types, we have effectively laid down the foundation of our proposed framework from which we will further extend its functionalities.

3.5 Separation of intent and implementation

In our proposed framework, there is a profound significance underlying intents as types. And this is because it allows for the separation of intents and implementations, which we believe to be critical towards enabling collaborations. Today, intent and implementation are intertwined which can be seen from how systems often specify and dictate how we carry out tasks in fulfilment of some intent. However, in context of the knowledge workplace, this is probably too draconian and rigid where the creativity and autonomy of individuals are especially valued. Furthermore, in reality, intents are often separate from implementations and may even be contributed by different people. Particularly, collaboration at scale is complex and we cannot reasonably expect it to be well-defined or pre-defined from the onset. On contrary, we can expect that for any collaboration:

- Multiple parties are involved.
- Coordination needs to happen vertically and horizontally within the organisation's hierarchy.
- People's ideas and ways of doing things can be fluid, dynamic and diverse.

Essentially, we need to flexibly handle the division of interdependent labour, interconnected intents, and diverse methods of handling the task at hand. Therefore, the question is: how can we tie people's collaboration together - managing the flow of information, etc. - allowing each person to define what they can do for others without overtly constraining their implementation, then allowing each task with input data to be broken into groups of data which can be handled with different implementations?

To achieve this, we believe that there needs to be a separation of intent from the implementation using type theory, enabled through our proposed framework.

3.6 Framework axioms

Next, we derive the rules (axioms) of our proposed framework which will govern the operations in a type theoretic manner.

Given some data $x : X$ of type X and an intent $G(x)$, the output is some implementation $g(x)$. We represent this statement type-theoretically in Eq. (4), which relates an implementation (term) to its intent (type).

$$g(x) : G(x) \tag{4}$$

Alternatively, we can write

$$g : \Pi_{x:X} G(x) \tag{5}$$

where we view g as a term of a dependent product type.

To fulfil an intent $G(x)$:

- there may exist data groups or subtypes $X_1, X_2, \ldots, X_k \subseteq X$ where the intent former G can be applied. This means that for different data $x_1 : X_1, x_2 : X_2, \ldots, x_k : X_k$, we could form different intents.

$$G(x_1), G(x_2), \ldots, G(xk) \tag{6}$$

- Given some data $x_i : X_i$, the intent $G(x_i)$ may have one or more implementations. Moreover, there may exists any number of possible strategies or implementation formers $g_1, g_2, \ldots, g_m$ for constructing implementations for $G(x_i)$.

$$g_1, g_2, \ldots, g_m : G(x_i) \tag{7}$$

- Each implementation former g_j for $G(x_i)$ may consume the outputs from one or more constituent intents $\Gamma_1, \Gamma_2, \ldots, \Gamma_n$ and associated implementations may receive inputs from one or more constituent intents (Eq. (8)). This means that an intent may contain its own hierarchy of constituent intents.

$$\begin{aligned} \gamma_1 : \Gamma_1(x_i) &\rightarrow \gamma_2 : \Gamma_2(x_i, \gamma_1) \\ &\rightarrow \ldots \\ &\rightarrow \gamma_n : \Gamma_n(x_i, \gamma_1, \ldots, \gamma_{n-1}) \\ &\rightarrow g_j(x_i, \gamma_1, \gamma_2, \ldots, \gamma_n) : G(x_i) \end{aligned} \tag{8}$$

In summary, these framework rules allow us to:

- Define and construct intents.
- Connect intent to constituent intents.
- Associate an intent with its implementation(s) and related data.

3.7 Group and assign

Next, to complete the framework, we will introduce two algorithms, Group and Assign, as abstract methods respectively described in Algorithm 1 and Algorithm 2.

The Group function (Eq. (9)) is defined as: For every intent, we have a set of data that can be further grouped into smaller groups based on some grouping criteria J.

$$Group(G, [x]) \rightarrow (G, [(x,j)]) \tag{9}$$

Algorithm 1 Group.

Input: G, $[x]$ where $x \in X$, J where $j \in J$
Output: $(G, [(x,j)])$
Initialise:
  $L \leftarrow \emptyset$ where L is a placeholder list to collate all (x,j) pairs
for each j in J **do**
  if x matches criteria j **then**
    $L \leftarrow (x,j)$
  end if
end for
$(G, [(x,j)]) = (G, L)$
return $(G, [(x,j)])$

The Assign function (Eq. (10:)) is defined as: For each group belonging to an intent G, some implementation g is defined and applied to the group.

$$Assign\left(G, \left[(x, j]\right) \rightarrow \left(G, \left[x, j, g_j(x, \gamma_1, \gamma_2, \dots, \gamma_m)\right]\right) \tag{10}$$

Algorithm 2 Assign.

Input: (G, [(x, j)])
Output: $\left(G, \left[\left(x, j, g_j\right)\right]\right)$
for each (x,j) in $[(x,j)]$ **do**
  if some g exists for (x,j) **then**
    $\left(x, j, g_j\right) \leftarrow (x,j)$.append g_j
  end if
end for
return $\left(G, \left[\left(x, j, g_j\right)\right]\right)$

Our proposed framework is collectively formed by the framework rules (Section 3.6), Group and Assign. This completes the description of our framework and in the following sections, we will discuss the evaluation strategies and findings for our proposed framework.

3.8 Evaluation approach

Our proposed work brings together type theory, type theoretic framework axioms and associated functionalities as a human AI intent orchestration framework intended for real world application. To the best of our knowledge, this is a novel effort and uniquely positioned idea to introduce capabiltiies for enabling a collaborative human AI future. This concurrently presents a challenge for us in determining how best to provide an evaluation of the proposed framework as widespread adoption and understanding will require more time and effort beyond our scope of research, as is reasonably expected given that the collaborative human AI society has yet to be a norm at the point of writing.

Going into the future, we intend to utilise our proposed framework and progress beyond our preliminary evaluation efforts to progressively identify and engage external parties for further evaluation through joint collaborations. Nevertheless, we endeavour to provide an evaluation of our proposed framework here and therefore consider the following:

- What is the closest and most relevant domain for our proposed framework?

- In reference to this domain, what are some useful evaluation strategies?

We believe that evaluation strategies for a toolkit (which we liken as comparable to our proposed framework) from the domain of human computer interaction [15] is relevant and suitable. We also note that the authors have expressed that "The problem is that toolkit evaluation is challenging, as it is often unclear what 'evaluating' a toolkit means and what methods are appropriate.", which speaks of a similar challenge for us and is still commonly faced till date by researchers of toolkits. Concretely, we reference and adopt two well-established evaluation strategies for the purpose of our evaluation, namely:

- Demonstration: As the name suggests, this evaluation strategy shows what the framework can support and how users might potentially use the framework through means such as using examples to illustrate a variety of possible applications. And it helps with the question of "What can be done with the framework". For our proposed work, we conduct this in the form of a "How To" scenario which is a technique of the demonstration evaluation strategy. Essentially, it is a walkthrough on a step-by-step breakdown of the workflow into individual steps and its associated results. Following this method, we demonstrate using a real-world context and step-by-step breakdown of how the framework orchestrates the intents of multiple users (Section 3.9).

- Usage: This particular evaluation strategy involves a user group in how they work with the framework, which helps with verifying aspects such as conceptual clarity, ease of use, value as perceived to the target user group, etc. And it helps with the question of "Who can use the framework". In practice, this is complementary and often combined with demonstrations. For our proposed work, we conduct this in the form of a walkthrough which is a technique of the usage evaluation strategy and gather the users' overall impressions. Using this method, we show our proposed framework to our potential users for their feedback and impressions. As a further enhancement, we also made this an interactive walkthrough where the potential users participated under the context of a work task in our team (Section 3.10).

3.9 "How to" scenario

To illustrate the walkthrough, we utilise our prototype software library and further implement a demo application built on top to illustrate plausibility and practical usability of our proposed framework. **Figures 3–6** are screenshots taken from the demo application.

To begin, let us suppose the scenario where we are planning for a fast food restaurant operations. Serving meals to our customers would be core to our business. In this context, "Serve Meal "would therefore be an overall intent from a management perspective. Another person on the team might look at this and suggest that we offer nuggets. Another then contributes that selling burgers will be a great idea too. Now, we have three intents altogether (**Figure 3**): "Serve Meal" and two constituent intents, "Serve Nugget" and "Serve Burger".

Subsequently, another team member points out that a meal would only be complete with a drink and further contributes "Serve Drink" (**Figure 4**). Here, we see that the framework is flexible to handle contributions of intents from different parties in a distributed manner. We could go on and add more intents, but this will suffice in this walkthrough for now.

At this point, we start to have a semblance of a plan on our food menu strategy and at work, this is what is often referred to as a "high level" view. However, it clearly still lacks further granularity such as:

- What type of burger?
- What type of drinks?
- What type of nuggets?
- What are the types of meal combinations we want to offer in our food menu?

We decide then that what we serve will depend on our inventory except we will not serve beef nuggets because it is not a norm. Looking through our inventory, we have chicken and beef in our raw meat inventory and coca-cola in our drinks. So, we will make beef burgers, chicken nuggets and coca-cola drinks. In doing this, we have effectively created groups of data based on some criteria and associated them with the corresponding implementation and intent (**Figure 5**).

Finally, we decide to offer 2 types of meal: Beef burger with coca-cola and chicken nuggets with coca-cola. This leads us to having a complete plan (**Figure 6**): We know what we want to do (Intent), we know how to do it (Implementation) and we have the ingredients (Data).

It is clear that different levels of details (i.e. intents, implementations) are often being handled by different workers horizontally as well as vertically within a company hierarchy. Any work operations of some scale will necessitate this division of labour. This is where disconnects will happen because there are no guarantees of higher level details connecting with more granular details, especially more so when the big picture is contributed by many different workers.

Figure 3.
Constructing intents in the context of menu planning for a fast food restaurant.

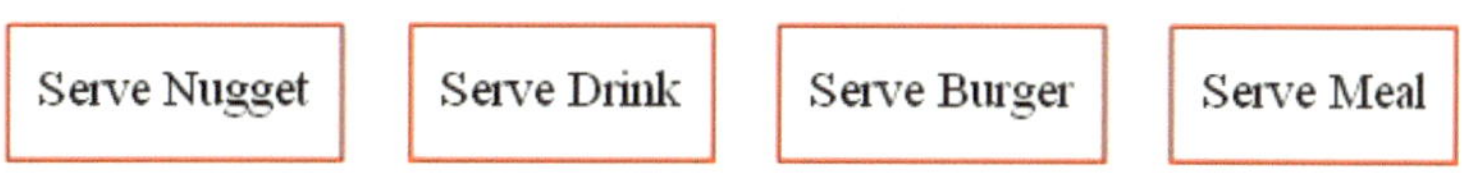

Figure 4.
Adding an intent "serve drink".

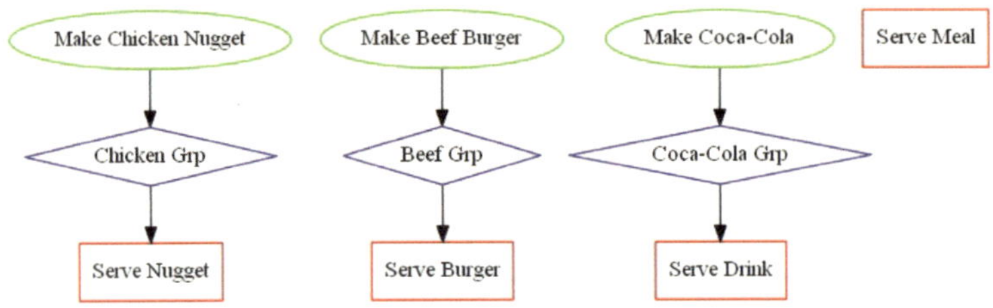

Figure 5.
Associating intent-data-implementation. Intents are represented as red rectangles, data is represented as blue diamonds and implementations are represented as green ovals.

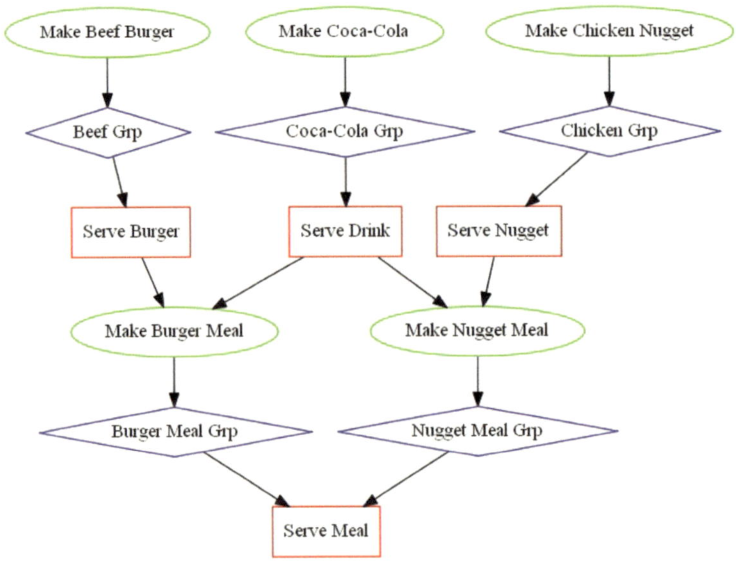

Figure 6.
Completed plan for offering fast food meals.

3.10 Usage feedback

To collect usage feedback, we conducted an interactive walkthrough on a work planning task for strategising our business unit growth, where we collaboratively develop a workplan by using our proposed framework principles. For confidentiality reasons, the actual content of the workplan will not be documented here. However, the process is similar in nature as our detailed step-by-step breakdown in Section 3.9. Following the conclusion of our interactive walkthrough, the user feedback for our proposed framework are summarised as follow:

- Structured yet flexible: The intent-data-implementation pattern for building up a workplan provided a clear structure for interdependent labour, interconnected intents, and diverse methods of handling the task at hand. However, there is still much flexibility for expressing intents and implementations, and the structure does not inhibit this flexibility.

- Clarity and precision: It is apparent from the workplan to see which intents are unfulfillable due to the inability to provide implementations or data for. While

DOI: http://dx.doi.org/10.5772/intechopen.96739

this feels rigid initially, it is subsequently well-accepted as there is a recognition that we cannot simply speak about our intents at work without the means to make it happen.

- In summary, we like to highlight some key points:

- The workplan can be easily visualised as a simple directed graph that is recursively constructed through primitive blocks comprising of 3 types of nodes: Intent, Implementation and Data.

- Beneath the apparent simplicity, the workplan is type theoretic and built upon the rules and algorithms (Group and Assign) described respectively in Sections 3.6 and 3.7. This confers any workplan built upon this framework with the desirable type theoretic properties discussed in Section 2.

From our evaluations, we demonstrate the plausibility of our proposed framework towards its intended goal for orchestrating work plans across a heterogeneous network of human intents associated with AI/human implementations and data.

4. Conclusions

In summary, we can see that the framework orchestrates the intents and associated implementations from different people while keeping the intent and implementation separate.

- This allows for each person to define what they can do for others in a distributed manner, while also enabling them the flexibility and freedom to provide their implementations they deem best.

- More so, by utilising primitive blocks of intent-data-implementation, the workplan is built up in a type theoretic manner where dynamic dependencies can be captured and represented clearly.

- This essentially makes the workplan a mathematical model which can be fully described using type theoretic expressions and hence is computable and constructive in nature.

4.1 Challenges and future directions

At the beginning of the chapter, we established the significance of human AI collaboration, and proceeded to share about our proposed work and its intended contributions towards this goal. In ending this chapter, it is apt to candidly discuss about its potential challenges and associated future directions. To do this, let us expand our view of human AI collaboration beyond the technological lens. Again, what then is human AI collaboration? It is a relationship, fundamentally. Like any successful relationship, trust and communication are crucial. Let us discuss each of these factors:

- Trust. This represents our confidence level in relying on the output from our AI counterpart. How do we ensure that the AI is performing as it should? Is there transparency in the way the AI operates? Are we able to interpret and understand why the AI does what it does? For example, standard-setting organisations define criteria for many technologies to ensure that compliance guarantees quality, digital security protocol such as SSL ensures the security for internet communications, well-documented manuals aid us in product troubleshooting and maintenance, etc. Every new technology introduced would eventually face questions such as these. Going forward, an area of potential interest lies in the framework integration with proof assistant capabilities. Work plans built upon our proposed framework are type theoretic in nature. The broad idea is therefore to treat every work plan as a theorem to be proven. By proving the theorem (work plan), the strong implication is that the work plan is verified to be working as intended. The ability to frame real world work plans as a mathematical model has desirable benefits in terms of trust, and this is an area which we hope to investigate more deeply.

- Communication. This represents how human and AI convey and exchange information. While our proposed work contributes towards a facet of human AI collaboration to enable the description and orchestration of intents across a network of humans and machines, it is targeted and focused as is the nature of research. As an analogy, while networking protocols enable the exchange of information over the Internet, it does not inherently make information easily searchable by users. For this, technologies such as search engine come into play. Switching back to our context here, an interesting aspect of communication (beyond our proposed work) would be to consider how these intents (along with its associated implementations and data) can be made discoverable and reusable by others.

Naturally, our discussion here is by no means exhaustive. It is our intent and hope that our proposed work and discussion contributes towards and catalyse future discussions in the research community for the continued advancements of human AI collaboration and ultimately, towards the future of a collaborative human AI society.

Acknowledgements

In making this work possible, I gratefully acknowledge the support of NVIDIA and research funding from Singapore Economic Development Board that enabled the opportunity for conducting this work.

Author details

Aik Beng Ng[1]*, Simon See[1], Zhangsheng Lai[1] and Shaowei Lin[2]

1 NVIDIA Corporation, San Tomas Expressway, Santa Clara, CA

2 Singapore University of Technology and Design

*Address all correspondence to: aikbengn@nvidia.com

References

[1] CNA. Meet the robot dog promoting safe distancing in Singapore's parks [Internet]. 2020. Available from: https://www.channelnewsasia.com/news/singapore/meet-the-robot-dog-promoting-safe-distancing-in-singapore-s-12716544 [Accessed: 2020-05-09]

[2] BBC. Japanese insurance firm replaces 34 staff with AI [Internet]. 2017. Available from: https://www.bbc.com/news/world-asia-38521403 [Accessed: 2017-01-05]

[3] MITTR. Our weird behavior during the pandemic is messing with AI models [Internet]. 2020. Available from: https://www.technologyreview.com/2020/05/11/1001563/covid-pandemic-broken-ai-machine-learning-amazon-retail-fraud-humans-in-the-loop/ [Accessed: 2020-05-11]

[4] Jie Ren and Peter J. Liu and Emily Fertig and Jasper Snoek and Ryan Poplin and Mark A. DePristo and Joshua V. Dillon and Balaji Lakshminarayanan. Likelihood Ratios for Out-of-Distribution Detection. In: Advances in Neural Information Processing Systems (NEURIPS '19); 2019. p. 14707–14718

[5] Cem Dilmegani. 995 experts opinion: AGI / singularity by 2060 [Internet]. 2021. Available from: https://research.aimultiple.com/artificial-general-intelligence-singularity-timing [Accessed: 2021-02-08]

[6] Federico Berruti and Pieter Nel and Rob Whiteman. An executive primer on artificial general intelligence [Internet]. 2021. Available from: https://www.mckinsey.com/business-functions/operations/our-insights/an-executive-primer-on-artificial-general-intelligence [Accessed: 2021-02-08]

[7] NIPS. From System 1 Deep Learning to System 2 Deep Learning [Internet]. 2019. Available from: https://nips.cc/Conferences/2019/ScheduleMultitrack?event=15488 [Accessed: 2019-12-11]

[8] Devlin, Jacob and Chang, Ming-Wei and Lee, Kenton and Toutanova, Kristina. BERT: Pre-training of Deep Bidirectional Transformers for Language Understanding. In: Proceedings of the 2019 Conference of the North American Chapter of the Association for Computational Linguistics: Human Language Technologies, Volume 1 (Long and Short Papers); 2019. p. 4171–4186

[9] T. Brown and B. Mann and Nick Ryder and Melanie Subbiah and J. Kaplan and Prafulla Dhariwal and Arvind Neelakantan and Pranav Shyam and Girish Sastry and Amanda Askell and Sandhini Agarwal and Ariel Herbert-Voss and G. Krüger and T. Henighan and R. Child and Aditya Ramesh and D. Ziegler and Jeffrey Wu and Clemens Winter and Christopher Hesse and Mark Chen and E. Sigler and Mateusz Litwin and Scott Gray and Benjamin Chess and J. Clark and Christopher Berner and Sam McCandlish and A. Radford and Ilya Sutskever and Dario Amodei. Language Models are Few-Shot Learners. In: 34th Conference on Neural Information Processing Systems (NeurIPS 2020), Vancouver, Canada; 2020.

[10] Sørensen, Morten Heine and Urzyczyn, Pawel. Lectures on the Curry-Howard isomorphism. 1st ed. Elsevier Science; 2006. 456 p. Hardcover ISBN: 9780444520777

[11] Gonthier, Georges. A Computer-Checked Proof of the Four Colour Theorem [thesis]. Microsoft Research Cambridge; 2005.

[12] Inria. COMPCERT [Internet]. 2008. Available from: http://compcert.inria.fr/ [Accessed: 2020-12-03]

[13] The Univalent Foundations Program. Homotopy Type Theory: Univalent Foundations of Mathematics [Internet]. 2020. Available from: https://homotopytypetheory.org/book [Accessed: 2020-12-03]

[14] Szabó, Zoltán Gendler. The Stanford Encylopedia of Philosophy [Internet]. 2017. Available from: https://plato.stanford.edu/archives/sum2017/entries/compositionality/ [Accessed: 2020-12-03]

[15] Ledo, David and Houben, Steven and Vermeulen, Jo and Marquardt, Nicolai and Oehlberg, Lora and Greenberg, Saul. Evaluation Strategies for HCI Toolkit Research. In: Proceedings of the 2018 CHI Conference on Human Factors in Computing Systems (CHI '18); 2018. p. 1–17

Chapter 5

AI-Powered Virtual Assistants in the Realms of Banking and Financial Services

Margherita Mori

Abstract

This chapter aims at providing a framework for analysis on evolutionary trends in finance that have to do with technological progress and especially with artificial intelligence (AI) applications. The starting point can be identified with a survey on how they have modified the business areas involving banking and financial services and on what can be expected – in terms of future strategic shifts and behavioral changes – on both the supply and the demand sides. The next step revolves around a wider and deeper investigation on the role that virtual assistants have started to – and are likely to further – play in the areas under scrutiny: special attention is requested upon the provision of enhanced customer service support, including conversational AI and sound branding; implications encompass developments that are on the cards, based upon digitalization as a must – not just an option – as shown by the Covid-19 pandemic. Conclusions allow to emphasize the significance, advancing features and value of this conceptual paper, as it leads to sort out best practices and success stories that are worth disseminating and replicating to benefit not only individuals and enterprises having direct interest in them, but society as a whole.

Keywords: AI applications, banking, conversational AI, digitalization, financial services, sound branding, virtual assistant

1. Introduction

Remarkable improvements in computer and telecommunication technology have fueled financial innovation worldwide in the last few decades and are key to most developments under way, that encompass institutional, product and process innovations: they deal with new types of financial firms (such as specialist credit card companies, electronic trading platforms and direct banks), new financial services (such as derivatives, asset-backed securities and foreign currency mortgages) and new ways of doing financial business (such as virtual, home and phone banking); in other words, all three pillars that the financial system is generally thought of being based on – namely: financial institutions, markets and products – have been positively affected, to the point that "many of the things that seemed so incredible 10 years ago are now foundational" [1]. Looking forward, the incredible pace of technology-driven change sounds promising for further progress, that may prove beneficial non only to financial institutions, but also to their counterparts across all economic sectors and geographical areas.

Within this framework, a major role needs to be assigned to artificial intelligence (AI), as a "multidisciplinary topic, where researches from multiple fields as neuroscience, computing science, cognitive sciences, exact sciences and different engineering areas converge" [2]. Related applications have gained momentum in the realms of banking and financial services, as well as in other industries, thus leading to state that getting involved in AI is a must, rather than just an option: reference points can be easily identified with "machines or systems that can perform complex tasks normally considered to require 'intelligence' and thus thought to be the preserve of humans" while the meaning of AI has been explained by evoking "a computer system that can sense, comprehend, act and learn"; as a result, it can be argued that "by enabling machines to interact more naturally – with their environment, with people and with data – the technology can extend the capabilities of both humans and machines far beyond what each can do on their own" [3], p. 3.

These thoughts pave the way for emphasizing the cross-cultural implications of AI, with its specific challenges and opportunities, and for discussing about it in terms of "systems endowed with the intellectual processes characteristic of humans, such as the ability to reason, discover meaning, generalize, or learn from past experience" [4]. Along this line of analysis, it comes natural to sort out most recent trends in this field, which sounds like an invitation to shed special light on AI-powered virtual assistants: they tend to be perceived as innovative tools that can help financial institutions and especially banks – as well as other enterprises – to provide customer service support and carry out administrative tasks, though the pertaining scope seems much wider; therefore, it is worth investigating how these digital assistants can be usefully resorted to and fully exploiting their potential, in sight of contributing to advances in the financial industry and particularly in the banking sector, due to its relevance and even prominence in many operating areas.

2. From "bricks" to "clicks"

To begin with, it must be accounted for the significant improvements in information technology that have been translated into new means of making banking and financial services available, including delivering them electronically: no surprise that e-finance has increasingly expanded, at a relatively fast pace, as financial institutions have quickly perceived the risk of becoming obsolete if this evolving trend would not be endorsed; a case in point has to do with the automated teller machine (ATM), that stands as a major form of an e-banking facility and that has enabled customers to get cash, make deposits, transfer funds from one account to another and perform other financial transactions all day long, without interacting with a human being. Not only the adoption of this facility has contributed to keep operating costs at relatively modest levels for banks, but more convenience has been provided to their customers.

Meanwhile, the drop in the cost of telecommunications has encouraged these financial intermediaries to develop other innovations, such as those that fall under the umbrella of home banking: it allows banks' customers to conduct many of their bank transactions without even leaving the comfort of home; in turn, banks have reaped benefits that stem from bearing substantially lower transaction costs compared to those implied by having customers come to their premises. The success of ATMs has been acting as a catalyst for the introduction and spread of other innovative facilities, including the automated banking machine, which can be described as a combination in one location of an ATM, an Internet connection to the bank's website and a telephone link to customer service [5], p. 500.

Further innovations in the financial industry – and notably in the home banking area – have been stimulated by the decline in the price of personal computers and the increase in their presence in households, thus laying the foundations for the virtual bank (also called digital-only and online-only bank) as a new type of banking institution: it delivers its services through the Internet or other forms of electronic channels – instead of conventional branches – and merely exists in cyberspace, which takes home banking one step further by enabling customers to have a full set of banking services available at home 24 hours a day; accordingly, the need for a physical location as the main vehicle to handle financial transactions has started to fade away – up to the point that many actual and potential bank customers do not even feel this need any more – and "clicks" have begun to replace "bricks" thanks to this evolutionary business model in the banking industry. However, pure Internet banks can hardly be conceived as the wave of the future, with a combination of "clicks" and "bricks" being expected to establish itself as the predominant format in the banking sector, whereby remote banking can usefully complement the banking services provided in line with traditional standards.

3. Progress in financial technology

Focusing on astonishing developments that have been recorded in financial technology – shortly fintech, a combination of finance and technology – proves rewarding to draw an updated picture of the global financial system, to be acknowledged as the largest industry in the world: as widespread evidence implies, most recent advances in this context have allowed financial institutions to satisfy the needs of their target markets in novel ways, that have shaken up the historically change-resistant banking sector, beyond expectations, and to serve potential customers that would otherwise populate the market segments consisting of the unbanked and underbanked [6]; thanks to progress in fintech, the financial industry has been experimenting with virtual banking, as well as – in more general terms – with automation, predictive analytics, new delivery platforms, blockchain and distributed ledger technology, to mention just a few innovations in the field under scrutiny. Promising areas that still call for keen attention involve mobile payments, digital currencies, peer-to-peer lending and marketplace lending, and underlying issues need to be more carefully addressed, that deal with the "the use of new technologies to solve regulatory and compliance requirements more effectively and efficiently" [7], p. 2 (or regtech) and with the recourse to innovative technology by supervisory agencies to support their activity (or suptech) [8].

As far as key players, most fintech innovations have been generated – and further fintech innovative solutions can be expected to emerge – outside the conventional financial and banking system. New applications developed by bright minds have been largely driven by non-bank entities, including venture capital-backed fintech start-ups and non-traditional providers of financial services that often focus their operations narrowly on a subset of the financial sphere of the economy: success stories abound in the area of digital payments, as fintech start-up companies have provided quick and convenient payment options, that encompass the adoption of e-wallets; they have been increasingly used for paying online purchases and for making person-to-person payments, thanks to intrinsic simplicity, not to mention strategies that have been designed to attract users by providing reward points, cash back and other exciting offers.

Very smart solutions that have been recently developed include those based on biometric sensors: their installation entails another step in the ATM innovation, as they are set to replace the need for carrying plastic cards and for remembering the pin to get access to a bank account through the facility at issue; biometric ATMs use palm or fingerprint sensors, eye recognition and integrated mobile applications to identify the account holder, thus eliminating mistakes in recognizing authorized customers and granting them access even if their card has been lost. Challenges ahead encompass voice biometrics, that has already unveiled its multifaceted advantages for financial institutions and their counterparts on a significant scale, and behavioral biometrics, that allows banks to look at how consumers behave (for instance, on a mobile app or website) rather than using physiological characteristics, like fingerprints.

4. Financial institutions and AI

With fintech being more and more widely adopted, a major impact in the financial sector has been generated by leveraging some of the latest innovations that involve AI: financial institutions have started to resort to it in order to transform the customer experience by enabling frictionless, 24/7 customer interactions while saving on costs; certain AI use cases have been increasingly disseminated within the financial industry, especially by banks, that are estimated to be offered the greater cost saving opportunities by front- and middle-office applications. A case in point has to do with the recourse to chatbots that have been used for a while, not only in the banking industry, and that allow to conduct an online chat conversation via text or text-to-speech, as a cost-effective alternative to direct contact with a live human agent.

Unquestionably, AI has the potential to upgrade bank management by making operations and processes faster and safer, with their efficiency set to increase as a result, which leads to consider the "intelligent" bank as a reliable candidate to become the rule – rather than remaining the exception – sooner than expected. Further progress can be foreseen in the financial industry, as AI applications are not just limited to retail banking: not only they are set to positively impact every office at banks, but these applications can support all financial services providers to completely redefine how they work, how they deliver innovative services and how they transform customer experiences; to stress this point, even though consideration tends to be focused on front-office operations, the back- and middle- offices of investment banks and other financial institutions can also benefit from AI, as shown by the full range of channels to get usefully involved in this innovative wave, encompassing front-office (conversational banking), middle-office (anti-fraud) and back-office (underwriting).

While the number of financial institutions that avail themselves of AI technology is on the rise, the ones that will achieve successful implementation are those that can develop comprehensive strategies: they can help to sense, comprehend, act and learn, therefore envisaging a system that can perceive the world around it, analyze and understand the information it receives, take actions based on that understanding and improve its own performance by learning from what happened; these strategies can help to drive growth in banking and financial services, in sight of a more sustainable and inclusive financial system, which our global village needs and deserves, now more than even, due to the troublesome and persisting effects of the Covid-19 health emergency. It is noticeable that several trends in digital engagement have accelerated during this pandemic and the gloomy picture to be confronted with should push financial institutions to take advantage of AI technology as "the foundation for new value propositions and distinctive customer experiences", to compete successfully and thrive [9].

5. Focus on virtual assistants

Virtual assistants can be considered the most conspicuous way in which AI has modernized so far – and can still upgrade – the financial system, adding to the human version of these assistants: as shown by the more and more intensive reliance on chatbots, they have attracted the attention of financial institutions, as well as of firms across a wide range of other economic sectors, and are being viewed as a key ingredient to create differentiation in today's increasingly crowded landscape; the underlying technology includes application programming interfaces that allow to analyze data, as well as web- and mobile-based user interfaces, and to deliver the necessary insights to the end customer. No wonder that forward-looking financial institutions have taken a leap of faith by investing in digital assistants to make "contextual insights" available to the right persons at the right time and through the preferred channels [10].

AI makes a huge difference between chatbots and virtual assistants who are typically self-employed workers specialized in providing administrative services to clients while operating outside their offices: these independent contractors usually work from a home office; access is granted to them to the necessary documents remotely, which acts as a stimulus to explore the potential of these assistants in the post-pandemic scenario. Since working from home has become more accepted for both workers and employers, the demand for skilled virtual assistants can be assumed to grow and new opportunities are surfacing for virtual assistants who are skilled in social media, content management, graphic design, blog writing, bookkeeping and web marketing [11].

To make a long story short, one of the advantages of hiring a virtual assistant is the flexibility to contract for just the services that each employer needs. Actually, this type of worker has become prominent, as small businesses and start-ups have embraced the trend to rely on virtual offices to keep costs low and firms of all sizes keep increasing their use of the Internet for their daily operations: being virtual assistants classified as self-employed workers, a company willing to take advantage of their services would not have to grant the same benefits it would be requested to provide its employees with; furthermore, since virtual assistants work offsite, they are expected to arrange and pay for their own toolkit (for instance, computer equipment, high-speed Internet connection and commonly used software programs) and it should not even be accounted for workspace, including a desk, at the company's office.

6. Digital assistants versus chatbots

The trend towards increasing digitalization has allowed "intelligent" versions of virtual assistants – known as Intelligent Virtual Assistants, shortly IVAs – to proliferate: they can be defined as digital personal software-based agents that assist us in performing our daily activities and are conceived as being "similar to personal human assistants that, let's say, take down notes during a meeting, remind us to tend to our 'to-do-lists,' or read messages and emails sent to us" [12]; for instance, these virtual assistants can help us to control and manage smart devices, that have become essential to operate in the areas of remote banking. Going into a few technical specifications, IVAs consist of "advanced conversational solutions – equipped with NLU (Natural Language Understanding), NLG (Natural Language Generation), and Deep Learning, that enables them to understand and retain context and have more productive conversations with users" [13].

Even though both chatbots and IVAs are ripened fruits of AI, a clear-cut distinction needs to be made between them, as a precondition for implementing them wisely and especially for taking full advantage of the added value that each of them can provide: chatbots may simulate a conversational experience to a certain extent but are ultimately constrained by having to work off a limited script, as they lack the ability to learn over time and to adapt to context; on the other hand, IVAs have the advanced capabilities to truly serve "assistants" to customers and can emulate human interaction while carrying out a wide variety of tasks to fulfill a user's requirements. Therefore, these two AI applications cannot be confused as one, as it often happens.

To all intents and purposes, chatbots are generally used as information acquisition interfaces, for instance to extract product details, whereas IVAs can assist in conducting business: if you ask chatbots for virtual assistance – for instance, to remind you of meetings, to manage your to-do lists and to take down notes – they get confused and tend to search for clarification by keeping asking the same questions; anyway, chatbots play a crucial role in customer service, as customers can usefully interact with them to satisfy specific needs, for example to gain product-related information or even book an appointment with the product manager. By contrast, IVAs utilize dynamic conversation flow techniques to "understand" human emotions, thus enriching communications with humans and hence covering a greater scope of action, which includes a wider range of tasks, such as those involving decision-making and e-commerce [14].

7. The potential of conversational AI

Despite the distinctive features outlined so far, both chatbots and IVAs are considered conversational interfaces, which organizations – including financial institutions – have recently started to actively and significantly deploy to automate their internal business processes. These applications provide incredible value, as they help to develop promising strategies that leverage AI, and are also impacting our personal lives to a remarkable extent: more and more frequently, customer service programs have been enriched by resorting to AI-powered software that makes "intelligent" customer – as well as employee – experiences available; to stress this point, it must be acknowledged that with conversational AI not only customers but also employees get the answers they need fast.

It's more than simple "if-then" logic, since conversational AI incorporates natural language to make human-to-machine conversations more like human-to-human ones: the outcome can be described in terms of increased customer engagement, continued trust and reliability in doing business, across all industries, and the ability to make the best thinkers and doers in any organization more productive; as a matter of fact, tech-savvy companies are building AI applications to augment business productivity, as well as to innovate business operations, with the ultimate goal being to help boost revenues. Therefore, more and more organizations are extending their efforts to identify additional areas to leverage AI and derive maximum value from it, by resorting to either chatbots or IVAs, as a viable alternative to utilizing both after identifying the right areas of application for each of them.

In general terms, conversational AI refers to technologies which users can talk to: these applications use large volumes of data, machine learning and natural language processing to help imitate human interactions, recognizing speech and text inputs, and translating their meanings across various languages; as far as the outcomes, the applications at issue help to build task-specific, channel-agnostic experiences by integrating data from various systems and channels (like SMS, Voice, WhatsApp and Facebook Messenger), and to retool teams for operational efficiency by automating the known and handling off the unknown. However, experts tend to label conversational AI's current applications as "weak AI", whereas "strong AI" should focus on a human-like consciousness that can perform a wider field of tasks and solve a broader range of problems [15].

8. Evidence from the financial system

With conversational AI being still considered in its infancy, it is even too easy to foresee tremendous progress that can translate into more cost-efficient solutions for many businesses, including financial institutions: focusing on banks, they have been reportedly slow in adjusting to new technologies, since managers have traditionally proven reluctant to abandon tried and tested systems for untested advancements, and by the way investing in technological progress involves huge amounts of money, which would make the risk of failure extra high; however, the transition to "conversational banking" has begun on a global scale, thus persuading banks to increasingly view chatbots and IVAs as new age contact center executives. Actually, these institutions have been pushed to mark digital transformation as a top priority as they have faced competition from fintech start-ups that have engaged in providing faster and more convenient options to their traditional customers in the last few years [16].

Another relevant factor must be identified with the health emergency that has been caused by the global spread of the novel coronavirus since 2019 and that has made an increased need for online services to surface and accelerate right afterwards: according to the United Nations Development Programme (UNDP), "while the pandemic demonstrates the immediate benefits of digital finance, the disruptive potential of digitalization in transforming finance is immense" [17] and positive effects can be even expected as a contribution to sustainable development, especially to the achievement of the Sustainable Development Goals in the 2030 Agenda that was adopted by the United Nations (UN) in 2015 [18]; therefore, not only mobile payment technologies have transformed mobile phones into financial tools for billions and billions of people, but going digital has been positively impacting – and can further upgrade – both supply- and demand-side drivers that interact to ultimately deploy AI to advancing promising areas in the financial industry, such as those involving cryptocurrencies, peer-to-peer lending and crowdfunding, to mention just a few of them. In line with valuable research work by the Organization for Economic Cooperation and Development (OECD), another innovative field to be closely scrutinized is populated by the so-called robo-advisors, that are computer programs designed to generate investment advice according to customer data and that tend to be utilized "as a cheap alternative to human wealth advisors" [19], p. 12.

Not to miss any opportunity, it is worth analyzing the financial system as a whole, beyond the boundaries of the banking sector, which leads to shed unprecedented light upon insurance companies. To support this view, even a quick look at most recent literature can be a source of useful insights to emphasize the potential of the wide range of AI use cases in the market segment that they make up: this almost 300-year old industry has been relatively slow to react to the disruption brought about by the digital age but the rapid pace of technological innovation and changing customer expectations in the last few years have contributed to substantial improvements, with insurtech start-ups playing a key role not only to put forth innovative AI applications in the industry under examination, but also to force traditional insurance players to follow suit; as a matter of fact, AI can be applied to the insurance value chain via a number of entry points, to encompass many areas (such as product development, marketing and sales, underwriting and risk-rating, claims management, robo-advisory, process improvements and recruitment, besides customer service) [20].

9. Success stories

Accordingly, success stories have unfolded in the insurance industry to learn from, in order to contribute to advances in the financial sphere of the economy, with their valuable repercussions in the real one not to be underestimated. A case study that showcases useful implications deals with Allianz Taiwan Life Insurance Co. Ltd.: it wanted a mobile assistant solution that could work across platforms to better serve customers; using IBM Cloud and IBM Watson Assistant, the company created an AI-powered virtual assistant that is described as being "smart, secure and almost human" and that was forged to field 80 percent of its most frequent customer requests, to provide "real help in real time" [21].

Turning to the banking industry, it is interesting to look at UBank, a digital-only bank established in 2008 and headquartered in Sydney, Australia, that has been able "to shrink time to market" by building a loan app virtual assistant on IBM Cloud platform: after consulting with an IBM Watson and Cloud Adoption Leadership team, this bank launched several initiatives, including RoboChat, a virtual assistant that incorporates advanced technology to support the bank's home loan application form online and particularly to help customers apply for home loans; to see the benefits of IBM Cloud technology at work, UBank and an IBM Garage team selected an initial use case, focusing on the bank's efforts to attract interest in its home loan offerings. Rather than relying solely on an email campaign, this bank built an app that plugs into Facebook and lets customers refer Facebook friends to the home loan program to be promoted and essentially RoboChat has been set up as an additional staff member providing a specific set of skills within this bank's current live chat capability [22].

Another revealing case study involves Banpro, that was founded in 1991 to support the social and economic development of Nicaragua and is part of the banking group Grupo Promerica, with nine operating banks throughout Central America. In an effort to scale exceptional always-on customer service, this bank launched Finn AI's virtual assistant technology in February 2018 and full functionality is being supported in Spanish, the primary language required to serve Banpro's Central American audience: within just one year, this virtual assistant has been able to complete 91% of chats without the need for a human customer service agent and to resolve 80% of customer queries, freeing up human customer service agents' time to deal with more complex customer inquiries; the virtual assistant at issue is not just handling queries from existing customers, as a great impression is being created on new prospects that engage with it and ultimately become new customers, with most common questions from them being about eligibility, product features and the application process [23].

10. Unprecedented challenges and opportunities

Success stories encourage to proceed with investing – money, as well as time and efforts – in technological innovation, not only in the financial industry. Challenges and opportunities ahead include the recourse to sound branding (also known as sonic branding, audio branding and acoustic branding) as a strategic tool for financial services companies to communicate with customers: by tradition, money has been a visual and physical entity but financial institutions are now getting involved in technological progress that should help them become recognizable audio brand entities as well; the mass adoption of smart speakers with voice assistants that are designed to enable audio-search, command and transactional capabilities has widened the spectrum of channels through which consumers can interact with brands and is pivoting service technology firmly in the direction of audio [24].

A recognizable and reassuring sounding brand that people can hear and easily associate with the services provided by a bank is likely to help build trust and engagement, to be undoubtedly considered relevant in the financial industry more than in any other sector. As a reinforcement, it can be argued that brand engagement is reportedly far stronger when audio is treated as an equal and essential aspect of the brand: therefore, it makes sense that quite a number of financial institutions are already harnessing the power of a well-designed sonic strategy to boost their brand; looking far beyond the solitary sonic logo, these institutions are creating holistic systems of branded sound and music that are flexible and anticipate proof for the technological advances of the future [25].

Among the frontrunners, HSBC launched its "sound identity" in 2019, a year after refreshing its visual brand identity to focus on its hexagons in a bid to make its brand more consistent: a bespoke piece of music was chosen to help people instantly recognize this bank and was proposed as the "next natural phase", with the marketing team cooperating with the digital team to get the audio in the bank's apps [26]; one of the major motivations was to reduce the fragmentation of HSBC's brand and the audio generated a brand score that could be used across multiple experiences, both online and offline, to create a universal brand identity through sound, at a time when consumers are increasingly busy and distracted [27]. By the way, recent developments have even enabled smart speakers to be adopted for voice-activated banking and as we march forward into a post-Covid era, that should be ever more screenless, the role of sound is set to become ever more important for the industry under scrutiny (and beyond).

11. Conclusions

To conclude, technological progress and changing consumer habits bring about unprecedented challenges that even lead to question how banking brands can retain trust while physical currencies tend to disappear and real, human interactions seem to increasingly belong to the past. At the same time, valuable opportunities keep emerging, that are worth taking: although there is less human contact, interactions can be more personable through tailoring the experience around the customer; as shown by the recourse to a brand's "hymn", the sound of this experience can play an important part both functionally and emotionally.

It's no secret that technology keeps evolving. For instance, IVAs hold extensive capabilities to help revolutionize banking: the critical focus is to identify the right areas to deploy these AI applications to, as well as to leverage chatbots; compared to IVAs, they are said to lack "understanding" of human emotions but chatbots that can gauge human sentiments are now being developed with the help of AI emotional intelligence.

All in all, AI can be considered a game changer in the financial arena, as well as in the real sphere of the economy, and also has the potential to contribute to the 2030 Agenda that was set up by the UN to provide a shared blueprint for partnerships for peace and prosperity for people and the planet: accordingly, a broad approach should be assumed to give due credit to the digital transformation that is spreading on a global scale and that preludes to creating inclusive digital economies as non-negotiable; factors to be further investigated range from technological advances that keep stimulating progress in the financial industry (especially in the delivery of banking and financial services) to the efforts under way to deploy AI to build the post-pandemic "new normal" as a stepping stone to a "new future". Anyway, a mental shift still stands as a precondition for meeting the challenges that raising the bar of intelligence over time implies and for taking the underlying opportunities.

Conflict of interest

The author declares no conflict of interest.

Author details

Margherita Mori
Department of Industrial and Information Engineering and of Economics, University of L'Aquila, L'Aquila, Italy

*Address all correspondence to: margherita.mori@univaq.it

References

[1] Briggs B, Buchholz S. Introduction, Tech trends 2019: Beyond the digital frontier. Hermitage: Deloitte Development LLC; 2019.

[2] Rabuñal Dopico J R, Dorado de la Calle J, Pazos Sierra A. Preface, Encyclopedia of Artificial Intelligence, vol. I. Hershey-London: Information SCI; 2009. DOI: 10.4018/978-1-59904-849-9

[3] Jubray E, Graham T, Ryan E. The Intelligent Bank - Redefining Banking with Artificial Intelligence. Dublin: Accenture; 2018.

[4] Copeland B J. Artificial Intelligence. Chicago; Encyclopaedia Britannica [Internet]. 2020. Available from: https://www.britannica.com/technology/artificial-intelligence/Reasoning [Accessed: 2020-11-29]

[5] Mishkin F S, Eakins S G. Financial Markets and Institutions. Global edition. 2012, Harlow: Pearson; 2012.

[6] Mori M. Banking Underserved Market Segments. Open Journal of Social Sciences. 2019; 7:506-517. DOI: 10.4236/jss.2019.73042

[7] Institute of International Finance. Regtech in Financial Services: Technology Solutions for Compliance and Reporting. Washington DC; March 2016.

[8] Broeders D, Prenio J. Innovative technology in financial supervision (suptech): The experience of early users. Basel: Bank for International Settlements, Financial Stability Institute; July 2018.

[9] Biswas S, Carson B, Chung V, Singh S, Thomas R. AI-bank of the future: Can banks meet the AI challenge? New York: McKinsey & Company; September 19, 2020.

[10] PwC India. Chatbot: The intelligent banking assistant [Internet]. 2020. Available from: https://www.pwc.in/consulting/financial-services/fintech/fintech-insights/chatbot-the-intelligent-banking-assistant.html [Accessed: 2020-12-12]

[11] Kenton W. Virtual Assistant. New York: Investopedia [Internet]. August 25, 2020. Available from: https://www.investopedia.com/terms/v/virtual-assistant.asp [Accessed: 2020-12-5]

[12] Joshi N. Yes, Chatbots and Virtual Assistants Are Different! Jersey City: Forbes [Internet]. December 23, 2018. Available from: https://www.forbes.com/sites/cognitiveworld/2018/12/23/yes-chatbots-and-virtual-assistants-are-different/?sh=181f32146d7d [Accessed: 2020-12-6]

[13] Chatbot vs IVA: 9 Ways to Tell the Difference. Mumbay: Haptik [Internet]. February 26, 2020. Available from: https://www.haptik.ai/blog/chatbot-vs-intelligent-virtual-assistant/ [Accessed: 2020-12-6]

[14] Misal D. What is the Difference Between a Chatbot and Virtual Assistant. Bengaluru: Analytics India Magazine [Internet]. September 7, 2018. Available from: https://analyticsindiamag.com/what-is-the-difference-between-a-chatbot-and-virtual-assistant/ [Accessed: 2020-12-9]

[15] IBM Cloud Education. Conversational AI. Armonk: IBM Cloud Learn Hub [Internet]. August 31, 2020. Available from: https://www.ibm.com/cloud/learn/conversational-ai#:~:text=Conversational%20artificial%20intelligence%20(AI)%20refers,which%20users%20can%20talk%20to. [Accessed: 2020-12-11]. DOI: 10.1109/icce.2017.7889324

[16] Team Kore.ai. Now, More Than Ever, Is Time for Banks to Adopt

Conversational AI with Vigour. Orlando: Kore.ai [Internet]. Available from: https://blog.kore.ai/now-more-than-ever-is-time-for-banks-to-adopt-conversational-ai-with-vigour [Accessed: 2020-12-11]. DOI: 10.7551/mitpress/12450.003.0009

[17] Digital finance, a lifeline during COVID-19 crisis, can deliver long-term financing of the Sustainable Development Goals. New York: UNDP [Internet]. August 26, 2020. Available from: https://www.undp.org/content/undp/en/home/news-centre/news/2020/Digital_finance_during_COVID19_can_deliver_longterm_financing_SDG.html [Accessed: 2020-12-11]

[18] People's Money: Harnessing Digitalization to Finance a Sustainable Future. Final Report by the UN Secretary General's Task Force on Digital Financing of the Sustainable Development Goals. New York: UN; 2020.

[19] Vives X. Digital Disruption in Banking and its Impact on Competition. Paris: OECD; 2020.

[20] The Ultimate Guide to Conversational AI in Insurance. Singapore: KeyReply [Internet]. Available from: https://www.keyreply.com/en/conversational-ai-insurance#:~:text=Insurance%20providers%20can%20also%20use,entire%20insurance%20purchase%20journey%20themselves [Accessed: 2020-12-11]

[21] Poser C. Allianz created an AI-powered virtual assistant. Armonk: IBM [Internet]. 2020. Available from: https://www.ibm.com/case-studies/allianz-taiwan-life-insurance/ [Accessed: 2020-12-12]

[22] UBank: Bringing digital banking capabilities to market faster with an IBM Cloud solution. Armonk: IBM [Internet]. 2017. Available from: https://www.ibm.com/case-studies/ubank [Accessed: 2020-12-12]

[23] Case Study Banpro: Banpro's Finn AI-Powered Virtual Assistant Handles 91% of All Customer Service Queries. Vancouver: Finn AI [Internet]. 2020. Available from: https://www.finn.ai/case-study/banpro-case-study/ [Accessed: 2020-12-12]

[24] Arnese M. The sound of money – how financial services companies are using audio branding to communicate with customers. Global Banking and Finance Review [Internet]. June 12, 2020. Available from: https://www.globalbankingandfinance.com/the-sound-of-money-how-financial-services-companies-are-using-audio-branding-to-communicate-with-customers/ [Accessed: 2020-12-12]. DOI: 10.3138/9781487533137-007

[25] Williamson R. How brands can utilize technology for sonic branding. Global Banking and Finance Review [Internet]. London: GBAF; August 27, 2020. Available from: https://www.globalbankingandfinance.com/how-brands-can-utilise-technology-for-sonic-branding/ [Accessed: 2020-12-12]

[26] Vizard S. 'It could have gone horribly wrong': HSBC launches 'sound identity' in next phase of global brand refresh. Marketing Week [Internet]. London: Xeim; January 23, 2019. Available from: https://www.marketingweek.com/hsbc-sound-identity-brand-refresh/ [Accessed: 2020-12-12]

[27] Sentence R. How HSBC refreshed its brand with a "universal" sound. London: Econsultancy [Internet]. April 1, 2019. Available from: https://econsultancy.com/hsbc-brand-refresh-universal-sound-audio/ [Accessed: 2020-12-12]

Chapter 6

Specific Wear Rate Modeling of Polytetraflouroethylene Composites via Artificial Neural Network (ANN) and Adaptive Neuro Fuzzy Inference System (ANFIS) Tools

Musa Alhaji Ibrahim, Yusuf Şahin, Auwal Ibrahim, Auwalu Yusuf Gidado and Mukhtar Nuhu Yahya

Abstract

Lately, artificial neural network (ANN) and adaptive neuro-fuzzy inference system (ANFIS) models have been recognized as potential and good tools for mathematical modeling of complex and nonlinear behavior of specific wear rate (SWR) of composite materials. In this study, modeling and prediction of specific wear rate of polytetraflouroethylene (PTFE) composites using FFNN and ANFIS models were examined. The performances of the models were compared with conventional multilinear regression (MLR) model. To establish the proper choice of input variables, a sensitivity analysis was performed to determine the most influential parameter on the SWR. The modeling and prediction performance results showed that FFNN and ANFIS models outperformed that of the MLR model by 45.36% and 45.80%, respectively. The sensitivity analysis findings revealed that the volume fraction of reinforcement and density of the composites and sliding distance were the most and more influential parameters, respectively. The goodness of fit of the ANN and ANFIS models was further checked using t-test at 5% level of significance and the results proved that ANN and ANFIS models are powerful and efficient tools in dealing with complex and nonlinear behavior of SWR of the PTFE composites.

Keywords: artificial neural network, adaptive neuro fuzzy inference system, multi linear regression, specific wear rate, PTFE reinforced composites

1. Introduction

In the study of tribology, highly nonlinear and very complex relationship exists. Specific wear rate of materials especially polymer matrix composites emanates from scores of intricate associations on both microscopic and macroscopic levels between surfaces which are in contact [1]. These associations depend upon tribological, geometrical as well as material behaviors of the contacting surfaces and the sliding conditions for example, temperature, type of contact, lubricating conditions, applied load, etc. [2]. Simulation of tribological properties usually deals with building of mathematical models extracted from practical data. The numbers of these models were obtained to simulate specific wear rate of materials under restricted conditions. Yet, no distinctive model was universalized to reveal the specific wear rate of polymer matrix composites.

Of recent soft computing techniques such as artificial neural networks (ANNs) and adaptive neuro fuzzy inference system (ANFIS) have emerged as potential and effective tools to model wear property of poly-based composites, owing to their abilities to learn from experimental data and generalize [3]. The pioneering studies of exploring the potentials of these soft computing methods especially ANN in the prediction of wear properties were carried out by Hutching et al. and Jones, Jensen and Fusaro [4, 5], respectively. Thereafter, many researchers applied the methods to analyze and predict the wear property polymer matrix composites under different test conditions and material compositions. In the physical experimentation of wear simulation, known material compositions and properties, experimental parameters are fed into the ANN and ANFIS models as inputs and the anticipated specific wear rate responses of the virtual scenario are computed. The fundamental advantage of ANN and ANFIS modeling in comparison to other modeling techniques are in their capabilities to provide accurate approximations or predictions when complexity and nonlinearity are involved at the same time. Complexity and nonlinearity cannot be handled by traditional curve fits [1]. More so, ANN and ANFIS models can effectively deal with these.

Velten, Reinicke and Friedrich [6] explored the potential of ANN when they predicted wear volume of short fiber reinforced polymeric composites. They found that with increase in the number of inputs the prediction quality of the ANN model was improved. Zhang, Friedrich and Velten [7] used multilayered feed forward neural network to predict the coefficient of friction and specific wear rate of short fiber reinforced polyamide. The results indicated a good agreement with experimental results. Jiang, Zhang and Friedrich [8] applied ANN model to predict both the wear and mechanical properties of polymer matrix composites. They established a 3D plots to investigate the properties of the materials based on the material constitutions and the experimental conditions. They reported that a well-trained ANN could model the wear and that the results of the model were in good agreement with the computed results. Aleksendric and Duboka [9] used ANN method to predict the automotive friction material features at room temperature. Five different types of friction materials were fabricated and experimented for the prediction purpose and the ANN was trained with five different learning algorithms. They found that each learning algorithm performed differently from one another but concluded that Bayesian regularization algorithm produced the best result with a single layer. Aleksendric and Duboka [10] applied the ANN to look into the possibilities of prediction wear property of friction composites at elevated temperature. They reported that ANN was effective in prediction the wear behavior of the materials as its results were in good agreement with the experimental ones. Jiang et al. [11] predicted wear and mechanical properties of polyamide composites, Varade and Kharde [12] predicted the wear behavior of PTFE glass-fiber reinforced composite using ANN and Taguchi technique. They found that ANN performed better than that of conventional Taguchi method.

Mesbahi, Semnani and Khorasani [13] employed adaptive neuro fuzzy inference system (ANFIS) to investigate the specific wear loss of PTFE, graphite short carbon fiber and nano-TiO2. They reported that ANFIS model performed better than ANN model. Jarrah, Al-Assaf and El Kadi [14] used ANFIS to model the fatigue property of unidirectional glass/fiber epoxy composite subjected to tension-tension and tension-compression conditions. They reported that the results of the ANFIS model were better when compared to those of ANN technique. Vassilopoulos and Bedi [15] applied ANFIS to model and predict the fatigue behavior of multidirectional laminate composite. They reported that about 50% of the data was adequate to model and predict the fatigue behavior of the composite and the results were in agreement with the actual data.

From above, it can be established that ANN and ANFIS models, hold great potentials and are promising tools in the modeling of complex and nonlinear wear behavior of polymer-based composites. The aim of this study is to model and predict the specific wear rate (SWR) of polytetraflouroethylene (PTFE) reinforced with glass, carbon and bronze fibers. The results of the ANN and ANFIS models were then compared with multilinear regression (MLR) model to affirm their superiority to traditional curve fit.

2. Methodology

ANN and ANFIS models have exhibited great power in describing complex, noisy and nonlinear phenomenon like specific wear rate. In this study, specific wear rate of PTFE composites was modeled and predicted using ANN, ANFIS and MLR models with density, volume fraction, sliding distance, sliding speed and load as inputs while specific wear rate as output. PTFE is a synthetic flouropolymer of tetrafluoroethylene that possesses superior characteristics due to its molecular structure consisting of fluorine and carbon. PTFE is hydrophobic and exhibits low wear resistance because of its soft nature making its suitable for use as a single material for practical application [16]. Glass fiber (GF) is a material consisting of several fine fibers of glass. GF is less brittle, less strong and cheaper than carbon. GF is compatible with most of the synthetic resin, does not rot and remain unaffected by the action of rodents and insects. Carbon fiber (CF) is composed of thin, strong crystalline filament of carbon and has a diameter of about 5–10 μm in diameter. It is very strong, stiff, and light; its strength is five times that of steel and twice as stiff. When CF is added to polymer, it improves the tribological property of the polymer [17]. Bronze fiber (BF) is a metal fiber that consists of 88% of copper and 12% of tin. It is hard and brittle. Its properties depend on the composition of the alloying tin.

A total of 63 specific wear rate experimental dataset was collected from the works conducted by [18, 19]. Some mechanical and physical properties of the materials are as shown in **Table 1**.

2.1 Artificial neural network (ANN)

ANN is a computational technique based on mimicking the function of the biological neurons [20]. Three properties are employed in differentiating various ANN models which are learning algorithms, transfer function as well as network

PTFE +Filler	Color	TS (MPa)	FS (%)	ρ (gcm^3)
Bronze fiber	Brown	18.0	165	3.90
Glass fiber	White	19.5	235	2.10
Carbon fiber	Black	13.5	87	2.25

Table 1.
Some physical and mechanical properties of the PTFE reinforced composites.

architecture [21]. The principal parts of ANN are the nodes or neuron which process the data and the interconnections that show the interconnection power connected to numeric weights [21, 22]. **Figure 1** shows the input, hidden and output layers of ANN architecture [23]. The fundamental structure of the neuron is as indicated in **Figure 2**. Each neuron receives input data, assigns weight $w_i$to the input data that indicates the connection power for that input data for each connection. Thereafter, a bias b_i value is added to the total addition of the input data and corresponding weights (u) in accordance with (Eq. (1)).:

$$u_i = \sum_{j=1}^{N} w_i x_j + b_i \quad (1)$$

where x_i is the input data, j is the jth data, w_i represents the weight, b_i shows the bias and N stands for the total number of the data points.

The summation is transformed into output with the aid of a transfer (an activation) function $F(u_i)$, generating a value referred to as the unit's "activation", as provided in the (Eq. (2)).

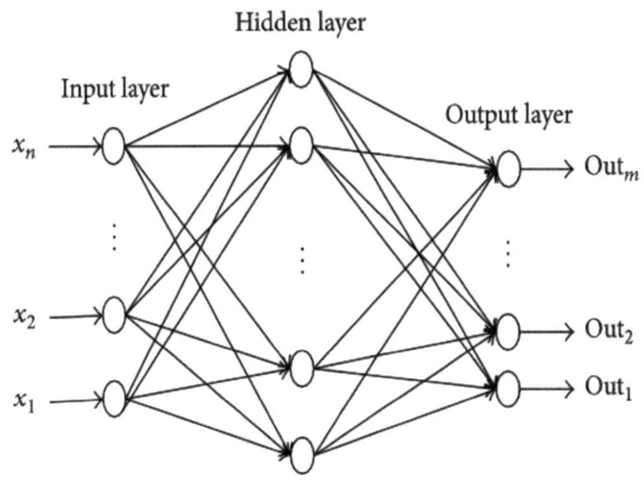

Figure 1.
A classical ANN image.

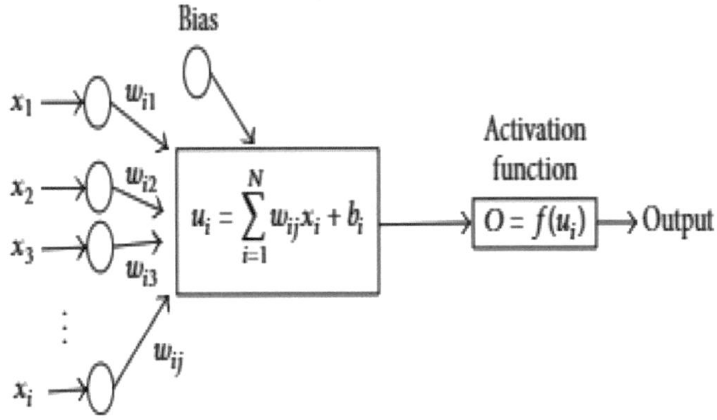

Figure 2.
The fundamental configuration of an artificial neuron.

$$O = f(u_i) \tag{2}$$

where O is the output.

One of the common types of ANN is the feed-forward neural network (FFNN). In FFNN technique, the processing layer is completely interrelated by weights to the rest of the processing layers (neurons). The learning stage in FFNN is actualized by back-propagation (BP) algorithm. The idea of using the BP algorithm is to compute the optimum weights that lead to the production of the target data in accordance with a chosen accuracy. In this paper, FFNN was applied due to its unique superiority of generating exclusive solutions without any prior knowledge of the mathematical computations in the parameters. **Figure 3** shows the architecturc of a FFNN used in this study. The ability of ANN to learn by example makes it suitable for solving complex and nonlinear behavior such as specific wear rate that cannot be addressed by conventional mathematical or physical models [24].

2.2 Adaptive neuro fuzzy inference system (ANFIS)

ANFIS is an important neurological network technique to obtain result of function approximation questions integrating the adaptive neural network and fuzzy inference system. As a global estimator, ANFIS was designed to surmount the limitations of FIZ and ANN. ANFIS integrates the experience capability of neurological network and the merits of the rule-based fuzzy structure, which can assimilate previous information into categorization mechanism. A structure is constructed by fuzzy logic descriptions as well as the neurological network is utilized to harmonize the structure variables naturally thus removing the demand for manual perfection of the fuzzy structure variables not like the neurological network where the structure is constructed by training. Adaptive ability and flexibleness of ANFIS makes it effective in handling the unpredictability of processes. The ANFIS architecture is made up of five different layers arranged like any multiple layer FFNN; coded in accordance with their operational functions. Sugeno firs-order fuzzy model had been applied in this paper. Different from ANN whereby weights are attuned, determination of the fuzzy language rules is needed as training the ANFIS model. The training of the membership function variables of the ANFIS is actualized through back propagation and/or least square and variables of the Takagi

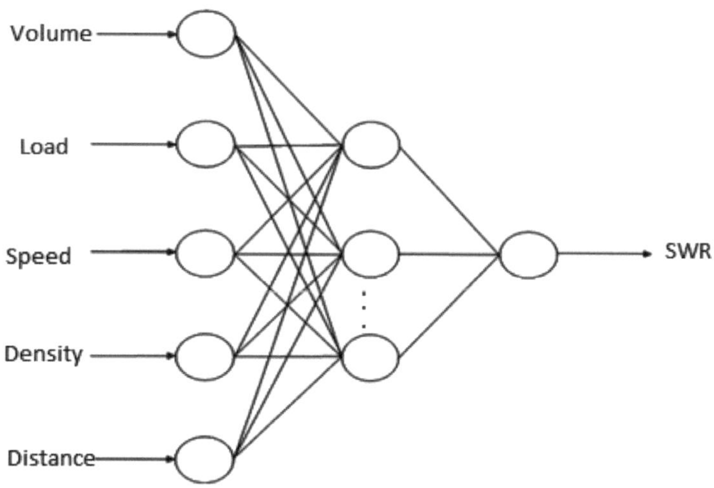

Figure 3.
ANFIS and first-order Sugeno FIS model configuration.

Sugeno fuzzy model are trained by the conventional square technique. The overall output of the ANFIS structure is described as a linear combination of the consequent variables. The common representation of an ANFIS model is demonstrated in **Figure 4** using two input variables.

Supposing fuzzy inference system with two inputs and one output as x, y and f, a Sugeno fuzzy first order, the rules are thus:

$$\text{Rule (1)}: \text{If } \mu(x)\ A_1 \text{ and } \mu(y) \text{ is } B_1; \text{then } f_1 = p_1x + q_1y + r_1 \quad (3)$$

$$\text{Rule (2)}: \text{If } \mu(x)\ A_2 \text{ and } \mu(y) \text{ is } B_2; \text{then } f_2 = p_2x + q_2y + r_2 \quad (4)$$

Membership functions parameters for x and y inputs are A_1,B_1, A_2, B_2 outlet functions' parameters of f are p_1, q_1, r_1, p_2, q_2, r_2, a five-layer neurological network arrangement possess the expression and configuration of ANFIS as:

First layer: Every node I is an adaptive node in this layer that contain the nodal function as:

$$\psi_i^1 = \mu_{Ai}(x) \text{ for } i = 1, 2 \text{ or } \psi_i^1 = \mu_{Bi}(x) \text{ for } i = 3, 4 \quad (5)$$

Where ψ_i^1 is for input x or y is the membership grade. Gaussian membership function had been selected in this paper because of its minimum prediction error.

Second layer: T-norm operator links every rule in this layer between inputs 'AND' operator thus:

$$\psi_i^2 = \beta_i = \mu_{Ai}(x) \times \mu_{Bi}(x) \text{ for } i = 1, 2 \quad (6)$$

Third layer: "Normalized firing strength" is the output of this layer:

$$\psi_i^3 = \varpi = \frac{W_i}{W_1 + W_2} \quad i = 1, 2 \quad (7)$$

Fourth layer: Each node i in the fourth layer is an adaptive node and executes the consequent of the rules as follows:

$$\psi_i^4 = \varpi(p_ix + q_iy + r_i) \quad (8)$$

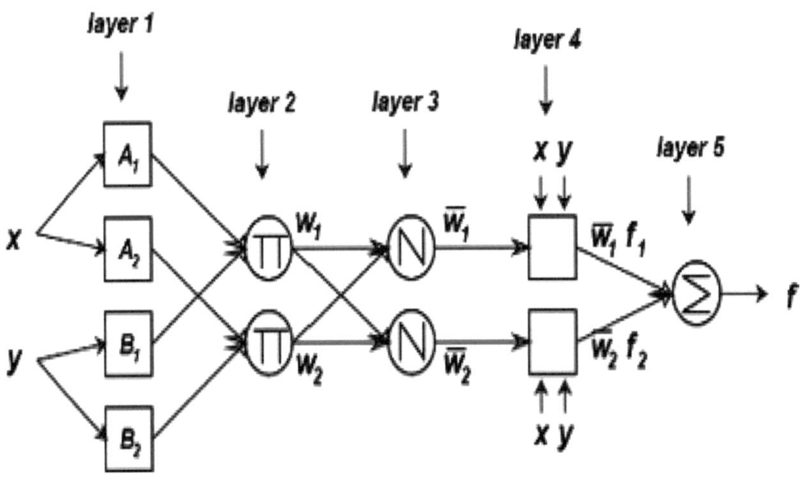

Figure 4.
ANFIS and first-order Sugeno FIS model configuration.

ϖ describes the output of layer 3 and p_i, q_i, $r_1$are the consequent parameters.
Layer 5: Here the overall output of all incoming signals is calculated in this layer as:

$$\psi_i^5 = \varpi(p_i x + q_i y + r_1) = \sum w_i f_i = \frac{\sum w_i f_i}{\sum w_i} \tag{9}$$

2.3 Multi linear regression (MLR) model

Linear regression analysis is a conventional technique used in applied science fields to describe and examine different parameters. Regression analysis especially aids in comprehending how the standard values of the dependent parameter varies as independent parameters vary, whilst the other independent parameters are held constant; examines the correlation between these parameters. The equation below was obtained from the regression analysis.

$$SWR = 0.162 + 0.269L + 0.369D - 0.293\rho + 0.347V + 0.0417S \tag{10}$$

where SWR is the (specific wear rate), L = applied load, D = sliding distance, ρ = density, V volume fraction of reinforcement and S = sliding speed.

2.3.1 Sensitivity analysis

In order to find the parameter that greatly influences the specific wear rate of the composites, nonlinear sensitivity analysis was conducted using neural network. In the sensitivity analysis each of the input parameter was used to predict the specific wear rate of the composites through the FFNN model. The performance of the individual model was assessed based on training and testing stages of the modeling. The mean value of the prediction performance criterion of each model obtained in both training and testing phases was then used to rank the contribution of the parameters to the specific wear rate of the composites.

2.3.2 Data pre-processing and performance evaluation

The data used in this study was normalized between zero (0) and unity (1) using the (Eq. (11)). The normalization was done to prevent bigger data values from overshadowing the smaller ones. Besides, data normalization simplifies the numerical computations in the model which in turn improves the prediction quality of the model and reduces the time taken to achieve global minimum.

$$\lambda_{norm} = \frac{\lambda - \lambda_{min}}{\lambda_{max} - \lambda_{min}} \tag{11}$$

Where λ_{norm} is the normalized mass loss value, λ_{min}, and λ_{max} represent actual, minimum and maximum mass loss values of the data, respectively.

The data was split into training data and testing data. The training data was used to adjust the weights of all the linking neurons until the required error level was attained. Consequently, the network performance is evaluated by using the testing data. The prediction performance is determined using Nush-Scutcliffe or determination coefficient (DC) and root mean square error (RMSE). DC indicates fitness of the observed data and lies between -∞ to 1 while RMSE measures the difference between actual and predicted values and ranges from 0 to 1. Higher B and lower RMSE indicate efficient model and vice versa. DC and RMSE are given in (Eq. (12)). and (Eq. (13))., respectively.

$$DC = 1 - \frac{\sum_{i=1}^{n}\left(\lambda_{acti} - \lambda_{predi}\right)^2}{\sum_{i=1}^{n}\left(\lambda_{acti} - \overline{\lambda}_{acti}\right)^2} \tag{12}$$

$$RMSE = \sqrt{\frac{\sum_{i=1}^{n}\left(\lambda_{acti} - \lambda_{predi}\right)^2}{N}} \tag{13}$$

Where N is number of observations, λ_{acti} stands for actual values, λ_{predi} represents the predicted values and $\overline{\lambda}_{acti}$ is the mean value of the actual values.

3. Results

3.1 Performances of the models

This section discusses the results obtained from the modeling of the study. The ANN, ANFIS and MLR models were also compared. **Table 2** showed the performance of the models.

3.2 Performance of the multilinear regression (MLR) model

In the MLR model, the data was split into two subclasses of training and testing. The ratios of the training and testing phases were characterized based on the fact that the common configuration of the model was built with respect to training data set. Hence, the quantity of data in the training category plays an important function. The total number of data was 63 in which 70% (44) and 30% (18) were randomly selected for training and testing, respectively. **Figure 5** shows the scatter plot of the relationship between actual and predicted specific wear rate (SWR) of the PTFE composites.

As it was shown in **Figure 5**, the determination coefficient (DC) of the training and testing phases were determined as 0.5674 and 0.5267, respectively. In addition, the RMSE in training was found to be 0.1275 but the testing stage RMSE was computed as 0.2306. As per the prediction analysis the DC and RMSE in the testing phase were considered. Therefore, MLR model with a DC of 0.5267 and RMSE of 0.2306 did not indicate higher prediction accuracy of the specific wear rate of the PTFE reinforced composites. This is attributed to the nonlinearity and complex nature of specific wear rate of the composites and MLR model is commonly good at finding linear and non-complex relationship between predictor and response variables [25].

3.3 Performance of the feed forward neural network (FFNN) model

Various learning algorithms were tried in order to find the optimum FFNN architecture and among all of them, Levenberg–Marquardt was found to be the

Model	Training		Validation		Testing	
	DC	RMSE	DC	RMSE	DC	RMSE
MLR	0.5674	0.1275	—	—	0.5266	0.2306
FFNN	0.9847	0.0341	0.9837	0.031	0.9749	0.0559
ANFIS	0.9847	0.0231	0.9956	0.025	0.9971	0.0168

Table 2.
Performance results of the models.

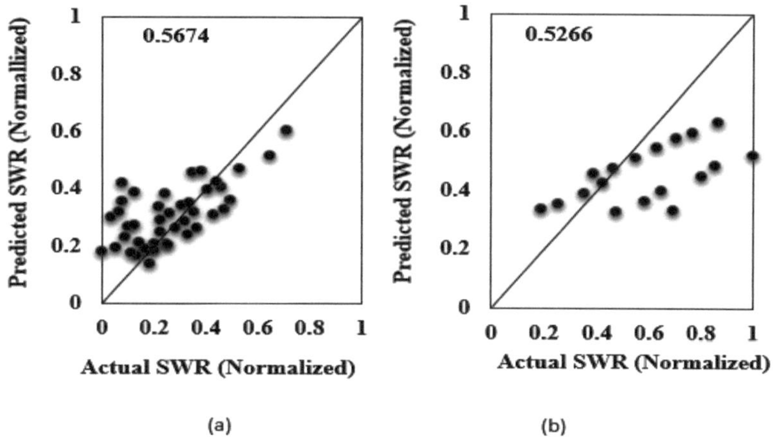

Figure 5.
Scatter plot of MLR model in (a) training and (b) testing stages.

most effective. In the FFNN model, the data is categorized into three subsets of training, validation and testing. The ratios of training, validation and testing are characterized based on the fact that fundamental architecture of the FFNN model is built based on the training data set. The whole data of the specific wear rate measurement was 63 in which 44 (70%) was chosen for training, 9 (15%) was selected for validation and 9 (15%) was chosen for testing. Besides, the sigmoid tangent was selected as the transfer function. The ANN model was trained with a single hidden layer. In addition, the number of neurons in the hidden layer was approximated using (Eq. (6)). Khademi et al. [26] instead of performing trial and error approach.

$$N_h \leq 2x_i + 1 \tag{14}$$

where N_h stands for the maximum number of neurons in the hidden layer and x_i equals the number of predictors. Therefore, in this research, based on the predictors which were five (5), the maximum number of neurons in the hidden layer was computed as eleven (11). The optimized ANN architecture with a single layer was thus expressed as [5–11-1]1.

Figure 6(a), **(b)**, and **(c)** shows the scatter plot of FFNN model in training, validation and testing stages, respectively. As seen in **Figure 6(a)** and **(b)** the FFNN model exhibited desirable results in both training and validation phases. Additionally, to estimate the prediction performance of the FFNN model, the DC was evaluated for the testing step as shown in the scatter plot of **Figure 6(c)**. As indicated in **Figure 6(c)**, the DC for testing of the FFNN model was determined as 0.9749 with a RMSE of 0.0559. This means that an FFNN model is more efficient in predicting the wear behavior of the composites, as compared to MLR model. This result was similar but higher than the previous study [27]. To round off, ANN model was found to be efficient in predicting the specific wear rate of the composites. This tallies with past studies of [28–29].

3.4 Performance of the adaptive neuro fuzzy inference system (ANFIS) model

In this study, ANFIS that used the hybrid learning algorithm was employed. The proportions training, validation and testing were chosen the same as the ones in FFNN modeling. To determine the best membership function, trial and error

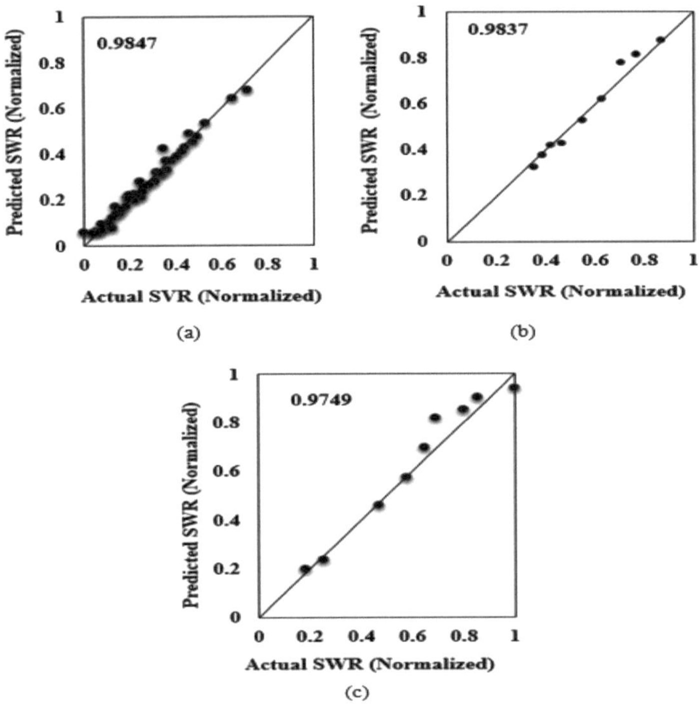

Figure 6.
Scatter plot of FFNN model in (a) training (b) validation and (c) testing phases.

approach was used and it was found that Gaussian membership function gave the best results at 50 epochs and 0.05 tolerance errors. **Figure** 7 shows the scatter plot of the relationship between the actual and predicted specific wear rate of the composites training, validation and testing stages. **Figure** 7 shows perfect coincidence of the target and the output data which demonstrated the capability of the ANFIS model. As it was indicated in the figure, the DC of the ANFIS in the testing stage was computed as 0.9971. More so, the RMSE was computed as 0.0225. To wrap up, ANFIS model was found to be capable of approximating the specific wear rate of the composites with satisfactory performance. This excellent performance of the ANFIS model agrees with the research by [30–31].

3.5 Comparing the results of FFNN, ANFIS and MLR models

In this article, the performance of FFNN, ANFIS and MLR models on predicting the specific wear rate of PTFE composites based on determination coefficient (DC) and root mean square error (RMSE) was investigated. The higher values of DC and lower values of RMSE indicate better and accurate prediction capability of model. For the purpose of the comparison, the data was split into 65% (40) and 35% (22) in training and testing, respectively for all the models. The comparative results of the models were shown in **Table 3** above. As seen in **Table 3**, the performances of the FFNN and ANFIS models were better than that of the MLR model. FFNN and ANFIS models outperformed the performance of the MLR model by 43.14% and 43.12% and 48.23% and 50.02% in training and testing phases, respectively. In other words, the prediction quality of MLR model was ineffective compared to the high prediction quality of ANN and ANFIS of 0.9783 and 0.9961, respectively. Their capabilities to predict the specific wear rate with minimum errors of 4% and 2%

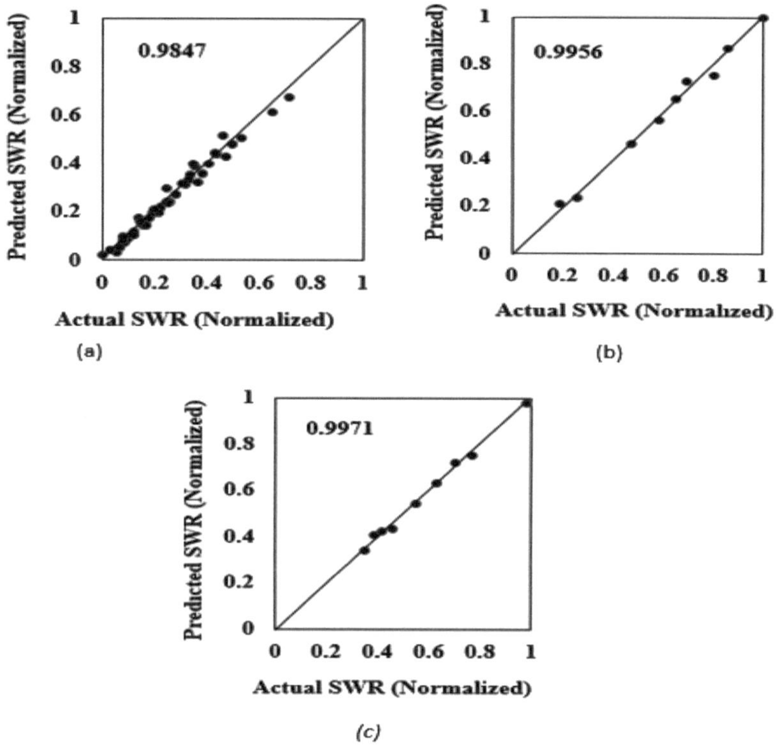

Figure 7.
Scatter plot of ANFIS model in (a) training (b) validation and (c) testing.

Model	Training		Testing	
	DC	RMSE	DC	RMSE
ANFIS	0.9841	0.0249	0.9961	0.0186
FFNN	0.9843	0.0248	0.9783	0.0441
MLR	0.5529	0.1314	0.4959	0.2067

Table 3.
Comparative performance results of the models.

(within acceptable level) as compared to the high error of MLR model of 21% is associated with their abilities to deal with nonlinear, noisy, complex relationship and to learn from outside environment and generalize. More so, the prediction performance of the ANFIS model was slightly higher than that of ANN model by 2%. This is because ANFIS model combines the attributes of both learning algorithm and fuzzy logic structure. **Figures 8** and **9** show the scatter plot of the models prediction quality and the simulated prediction results, respectively. It can be seen that ANFIS and FFNN models indicated perfect match with the actual SWR of composites while MLR model exhibited imperfect consistency with respect to the observed SWR of the composites.

3.6 Sensitivity analysis

Identification of most influential parameter in the study of wear is a significant step in achieving optimum results. In the light of this, a nonlinear FFNN sensitivity

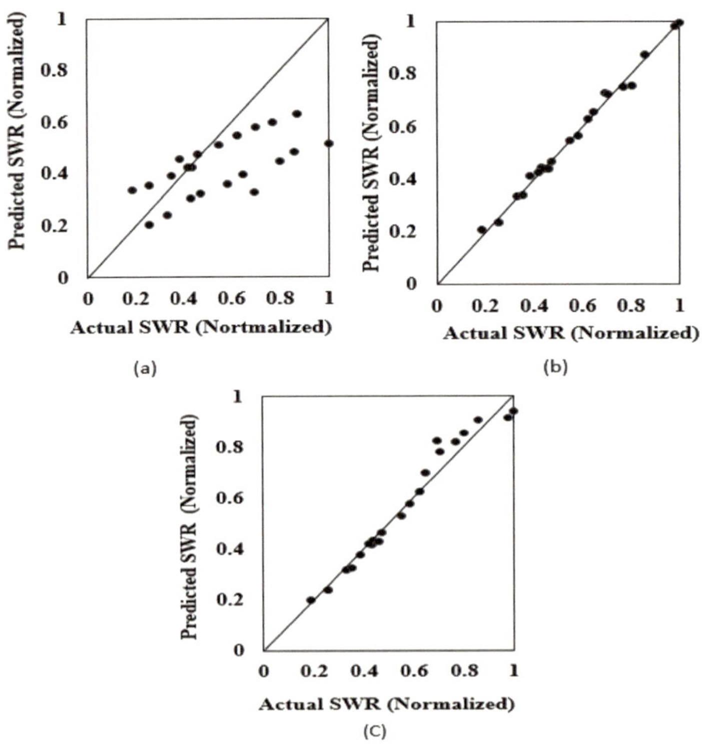

Figure 8.
Scatter plot of (a) MLR, (b) ANFIS and (c) FFNN models in testing stages.

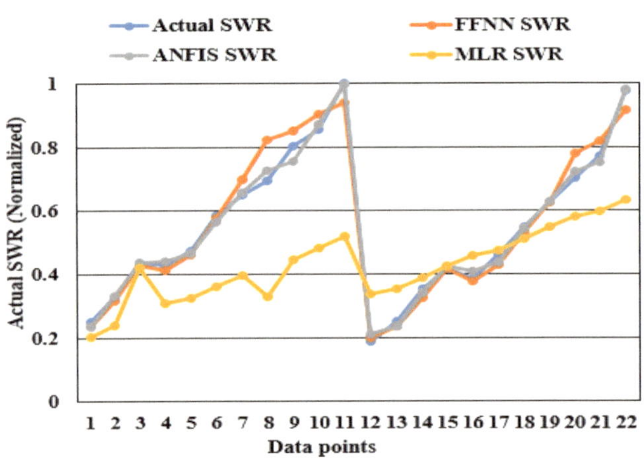

Figure 9.
Comparing the performance of the models: Testing stage.

of the specific wear rate of the composites was applied in this study to establish the dominant parameters in place of using traditional linear methods. The five specific wear rate were evaluated and ranked based on the mean value of the DC of the single modeling obtained in training and testing phases of the FFNN modeling. The results of the ranking based on the sensitivity analysis of the specific wear rate was presented in **Table 4**.

As seen from **Table 4**, in terms of the experimental conditions sliding distance is the most influential parameter, then sliding speed and the least was the applied load. On the contrary [27] reported that the sliding speed had the greatest effect on

Parameter	Average DC	Rank
Volume fraction	0.4658	1
Density	0.4027	2
Sliding distance	0.3503	3
Sliding speed	0.2985	4
Applied load	0.1476	5

Table 4.
Sensitivity analysis results of each input parameter.

Output ANFIS Model		FFNN Model		MLR Model	
SWR t-stat	t-critical	t-stat	t-critical	t-stat	t-critical
0.3464	1.6702	−0.4492	0.3464	0.4701	1.6702

Table 5.
Results of t-test at 5% significance level.

the volume loss of the polymer composites. This means that the various sliding distances can lead to different specific wear rate of the composites. The higher the applied load the more the composites will spend in the elastic deformation phases. With respect to the composites constitutions, volume fraction of the reinforcements had the greatest effect on the specific wear rate followed by density. This implies that as the volume fraction of the reinforcing phases was increased hardness with a corresponding increase in density that minimizes the specific wear rate of the composites. This agrees with the work of [13]. However, when all the parameters are compared it was found that volume fraction was the most influential and applied load presented the least effect on the specific wear rate of the composites.

The model's goodness of fit versus the actual values for the ANN and ANFIS models was tested using t-test at 5% level of significance and the outcomes revealed that there was no significance difference between the predicted and the actual values of the SWR. This was as shown in **Figures 6** and 7 and the t-test result was presented in **Table 5**.

4. Conclusions

In this study, three various data driven models namely: feed forward neural network (FFNN), adaptive neuro fuzzy inference system (ANFIS) and multi linear regression (MLR) were applied in modeling and prediction of the specific wear rate (SWR) of polytetraflouroethylene (PFTE) composites. MLR model with DC of 0.5266 and RMSE of 0.2306 was found to be inefficient enough to predict the SWR of the composites. This is due to the complex and nonlinear relationship between the investigated variables and MLR model is usually good at establishing linear relationship between predictors and responses. FFNN model having DC equals 0.9802 and RMSE as 0.0471 was found to be capable in predicting the SWR of the PTFE reinforced composites. ANFIS model DC equal to 0.9967 was found to be talented in approximating the SWR of the composites. FFNN and ANFIS models were found to be highly qualitative in predicting the SWR of the composites, yet MLR model was found to be incapable in the same prediction scenario. The high prediction performance of the FFNN and ANFIS models is owing to their capability to deal with nonlinear, noisy and complex relationship which is typical of SWR of the polymer composites. Although, both ANFIS and FFNN models were capable of predicting the SWR of the composites, ANFIS was found to be more efficient in predicting the SWR of the composites than FFNN model. The sensitivity analysis of the built FFNN model indicated that sliding distance was the dominant parameter on the SWR of the composites in terms of the experimental conditions while volume fraction of the reinforcing phases was also influential parameter on the SWR with respect to the composites compositions. However, considering all the input parameters volume fraction of the reinforcements was the most dominant parameter and applied load was the least parameter influencing the SWR of the PTFE composites. The goodness of fit was rechecked using t-test at 5% significance level and the results affirmed the superiority of the FFNN and ANFIS models as powerful and efficient tools of modeling and prediction of SWR of the PTFE composites.

Acknowledgements

The authors are grateful to all those who have assisted in this work.

Conflict of interest

The authors declare no conflict of any kind.

Author details

Musa Alhaji Ibrahim[1*], Yusuf Şahin[2], Auwal Ibrahim[1], Auwalu Yusuf Gidado[1] and Mukhtar Nuhu Yahya[3]

1 Kano University of Science and Technology, Kano, Nigeria

2 Mechanical Engineering in Nişantaşı University, İstanbul, Turkey

3 Bayero University, Kano, Nigeria

*Address all correspondence to: musaibrahim@kustwudil.edu.ng

References

[1] Abdelbary A., Abouelwafa M.N., El Fahham, I.M. and Hamdy, A. H."Modeling the wear of polyamide 66 using artifcial neural network". Material Design 41, (2012): 460–469.

[2] Gyurova Lada A., Paz Miniño-Justel, Alois K. Schlarb. "Modeling the sliding wear and friction properties of polyphenylene sulfide composites using artifcial neural network," Wear 268, (2010):708–714.

[3] Laurentiu Frangu and Ripa Minodora. "Artificial Neural Networks Applications in Tribology – A Survey". (2001).

[4] Rutherford K. L., Hatto P. W., Davies C., and Hutchings, I. M. "Abrasive wear resistance of TiN/NbN multi-layers: Measurement and neural network modelling". Surface and Coatings Technology 86–87, no. PART 2 (1996): 472–479.

[5] Jones Steven, P., Ralph Jansen, and Fusaro L. Robert. "Preliminary investigation of neural network techniques to predict tribological properties". Tribology Transaction 40, no. 2 (1997): 312–320.

[6] Velten K., Reinicke R., and Friedrich, K. "Wear volume prediction with artificial neural networks". Tribolology International 33, (2000): 731–736.

[7] Zhang Z., Friedrich k., and Velten K.. "Prediction on tribological properties of short fibre composites using artificial neural networks". Wear 252, (2002): 668–675.

[8] Jiang Z., Zhang Z., and Friedrich K. "Prediction on wear properties of polymer composites with artificial neural networks". Composites Science and Technology 67, no. 2 (2007): 168–176.

[9] Aleksendric D. and Duboka Č. "Prediction of automotive friction material characteristics using artificial neural networks-cold performance". Wear 261, (2006):269–282.

[10] Aleksendric D. and Duboka Č. "Fade performance prediction of automotive friction materials by means of artificial neural networks". Wear 262, no. 7–8 (2007):778–790.

[11] Jiang Z., Gyurova L., Zhang Z., Friedrich K., and Schlarb A. K. "Neural network based prediction on mechanical and wear properties of short fibers reinforced polyamide composites". Materials and Design 29, no. 3 (2008): 628–637.

[12] Varade B. V. and Kharde Y. R. "Prediction of specific wear rate of glass filled PTFE composites by artificial neural networks and Taguchi approach". International Journal of Engineering Research and Appllication 2, no. 6 (2012): 679–683.

[13] Haghighat A., Semnani D., and Nouri, S. "Performance prediction of a specific wear rate in epoxy nanocomposites with various composition content of polytetrafluoroethylen (PTFE), graphite, short carbon fibers (CF) and nano-TiO2 using adaptive neuro-fuzzy inference system (ANFIS)". Compos. Part B 43, no. 2(2012): 549–558.

[14] Jarrah M.A., Al-Assaf Y. and El-kadi H. "Neuro-fuzzy modeling of fatigue life prediction of unidirectional glass fiber epoxy composite laminates". Journal of Composite Materials 36, no. 06 (2001):685–700.

[15] Vassilopoulos A.P. and Bedi R. "Adaptive neuro-fuzzy inference system in modelling fatigue life of multidirectional composite laminates". Computational Materials Science 43, no. 4 (2008):1086–1093.

[16] Zhang Z., Breidt C., Chang L., Haupert F., and Friedrich K. "Enhancement of the wear resistance of epoxy: Short carbon fibre, graphite, ptfe and nano-tio2". Composite Part A Applied Science and Manufacturing 35, no. 12 (2004): 1385–1392.

[17] Li J. and Xia Y. C."The reinforcement effect of carbon fiber on the friction and wear properties of carbon fiber reinforced PA6 composites". Fibers Polymer 10, no. 4 (2009):519–525.

[18] Şahin Y. and Mirzayev H. "Wear characteristics of polymer-based composites". Mechanics of Composite Materials 51, no. 5 (2015):543–554.

[19] Unal H., Mimaroglu A., Kadiaglu U and Ekiz H. "Sliding friction and wear behaviour of polytetrafluoroethylene and its composites under dry conditions". Materials and Design 25, (2004):239–245.

[20] Parveen R., Nabi M., Memon F. A., Zaman S., and Ali M."A Review and Survey of Artificial Neural Network in Medical Science". Journal of Advanced Research in Computing and Applications 3, no. 1 (2016): 8–17.

[21] Mokhtari M. and Behnia M. "Comparison of LLNF, ANN, and COA-ANN techniques in modeling the uniaxial compressive strength and static Young's Modulus of limestone of the Dalan formation". Natural Resources Research 28, no. 1(2019): 223–239.

[22] Demuth H. and M Beale. Neural Network Toolbox For Use with MATLAB. 2004.

[23] Yurdakul M. and Akdas H. "Modeling uniaxial compressive strength of building stones using non-destructive test results as a neural networks input parameters". Construction and Building Materials 47, October 2013 (2013):1010–1019.

[24] Shebani A. and Iwnicki S. "Prediction of wheel and rail wear under different contact conditions using artificial neural networks". Wear 406–407, no. March 2017 (2018): 173–184.

[25] Tugˇrul Seyhan A, Gökmen Tayfur, Murat Karakurt and MetinTanogˇlu. "Artificial neural network (ANN) prediction of compressive strength of VARTM processed polymer composites". Computational Materials Science 34, (2004): 99–105.

[26] Faezehossadat Khademi, Sayed Mohammadmehdi Jamal, Neela Desphande, and Shreenivas Londhe. "Predicting strength of recycled aggregate concrete using artificial neural network, adaptive neuro-fuzzy inference system and multi linear regression". International Journal of Sustainable Built Environment 5, (2016):355–369.

[27] Halil Ibrahim, Kurt and Oduncuoglu Murat. "Application of a neural network model for prediction of wear properties of ultrahigh molecular weight polyethylene composites". International Journal of Polymer Science 2015, (2015): 1–11.

[28] Zhang Z., Barkoula N. M., Karger-Kocsis J., and Friedrich K. "Artificial neural network predictions on erosive wear of polymers,". Wear 255, no. 1–6 (2003):708–713.

[29] Aleksendrić D. and Barton, D. C. "Neural network prediction of disc brake performance". Tribology International 42, no.7, (2009):1074–1080.

[30] Haghighat Mesbahi A., Semnani D., and Nouri Khorasani S."Performance prediction of a specific wear rate in epoxy nanocomposites with various composition content of polytetrafluoroethylen (PTFE), graphite, short carbon fibers (CF) and nano-TiO2 using adaptive neuro-fuzzy

inference system (ANFIS)". Compososites: Part B 43, no. 2 (2012.): 549–558,

[31] Gupta Vikrant, Rama Singh, Jha Kumar Manoj, and Qureshi M.F."ANFIS prediction of the polymer and polymer composite properties and its optimization technique". AMSE JOURNALS-2014-Series Model. A 87, no. 2 (2014): 70–80.

Chapter 7

Intelligent Decision Support System

Moruf Akin Adebowale

Abstract

A phishing attack is one of the most common forms of cybercrime worldwide. In recent years, phishing attacks have continued to escalate in severity, frequency and impact. Globally, the attacks cause billions of dollars of losses each year. Cybercriminals use phishing for various illicit activities such as personal identity theft and fraud, and to perpetrate sophisticated corporate-level attacks against financial institutions, healthcare providers, government agencies and businesses. Several solutions using various methodologies have been proposed in the literature to counter web-phishing threats. This research work adopts a novel strategy to the detection and prevention of website phishing attacks, with a practical implementation through development towards a browser toolbar add-in. The IPDS is shown to be highly effective both in the detection of phishing attacks and in the identification of fake websites. Experimental results show that approach using the CNN + LSTM has a 93.28% accuracy with an average detection time of 25 seconds, whilst the approach has a slightly lower accuracy. These times are within typical times for loading a web page which makes toolbar integration into a browser a practical option for website phishing detection in real time. The results of this development are compared with previous work and demonstrate both better or similar detection performance. This is the first work that considers how best to integrate images, text and frames in a hybrid feature-based solution for a phishing detection scheme.

Keywords: cybercrime, deep learning, convolutional neural network (CNN), long short-term memory (LSTM), big data

1. Introduction

The use of technology for fraudulent activities has flourished in recent years. The technical resources required to carry out phishing attacks are readily available through private and public sources. Hence, some of these technical resources have been automated and streamlined, thereby allowing their use by non-technical criminals. This automation has made it easier for a larger population of less-sophisticated criminals to commit crimes online, as it has made phishing more viable and economical.

In the recent times, there has been a considerable increase in the assortment, technology and complexity of phishing attacks in response to the increase in countermeasures and user awareness in order to sustain profitability from the illegal activities by the phisher [1]. Providing the ability to detect website phishing attacks may help individual users or organisations in identifying legitimate websites. The effectiveness in recognising an attack may significantly contribute to the making of an effective decision between a fake and legitimate site [2]. Phishing is a form of social engineering attack in which an attacker, also known as a phisher, attempts to fraudulently retrieve sensitive user information by sending an email claiming to be a legitimately established organisation. They scam the user into giving confidential information that will be used for identity theft [2]. A phisher uses various methods, including email, web pages, and malicious software, to steal personal information and account credentials [3]. The aim of the phishing website is to use users' private information without their permission, and they do this by developing a new website that mimics a reliable website [4].

Hence, phishing website detection has become the object of a great deal of consideration among many academics who are attempting to find ways to incorporate malicious detection devices into web servers as a safety precaution [5]. Despite there being several ways to carry out phishing attacks, current phishing detection techniques unfortunately only cover some attack vectors such as fake website and emails [6]. Moreover, phishing has become more sophisticated, and such attacks can now bypass the filters that have been put in place by anti-phishing techniques [7]. Some detection techniques have been proposed, but most of them only deal with spoof web pages [8]. However, it is quite challenging in detection due to the evading techniques that the phisher uses.

Currently, machine learning is continuously demonstrating its effectiveness in an extensive range of applications. This technology has come to the fore in recent times, owing to the advent of big data [9]. Big data has enabled machine learning algorithms to discover more fine-grained patterns and to make more accurate and timely predictions than ever before [10]. Machine learning techniques are used for object identification in images, the transcription of voice into text, matching news items and products with user interests and presenting relevant search results [11]. The most common form of machine learning, whether deep or not, is supervised learning [12]. Previous methods have failed to combine the usage of frames, images, and text to develop an effective phishing detection method. Because using only text which is the common trend to a detection phishing website, this will not be effective as some changes can be made to the frame and the image. Doing so is, therefore, the focus of this work and therein lies its originality as well using the deep learning of Convolutional Neural Network (CNN) and Long short-term memory (LSTM) as classification algorithm in this solution.

Given the above, the objective is to develop a solution that includes the decision support system for detection of phishing attacks as well as providing insights and improving awareness as to how active Internet users can protect themselves against phishing attacks. It is hoped that this will help to formulate an upward trend in the practice of preventive measures against cyber-security issues. Despite various approaches having been utilised to develop anti-phishing tools to combat phishing attacks, these methods suffer from limited accuracy [1].

The main aim of this research is to develop an intelligent phishing detection and protection scheme for identification of website-based phishing attacks. This goal involves improving on previous work by building a robust classifier for intelligent phishing detection in online transactions. In order to achieve this aim the intelligent phishing detection support system should possess the following characteristics:

1. **Robustness**: It should have a hybrid algorithm that can support efficient classification for website phishing detection in real-time.

2. **Accuracy**: It should improve accuracy by reducing the false positive (FP) rate and increasing the true positive (TP) rate with absolute precision.

3. **Optimisation:** It should be able to optimise performance by employing a hybrid method that uses the features of website images, frames, and text for the user's objectives.

4. **Real-time functionality:** It should notify the user about the about the legitimacy of the website before the user web browser loads the intended page.

These requirements will be met by achieving the following five specific objectives:

I. Examine the Adaptive Neuro-fuzzy Inference System (ANFIS) algorithm as a baseline and the use of more advanced methods to improve accuracy.

II. Develop an algorithm that improves phishing-detection accuracy by comparing the text, images and frames of a given website with a knowledge model.

III. Train, test, and validate the developed system (machine learning) for real-time phishing detection.

IV. Automate the detection mechanism in real-time and test it offline.

V. Develop a plug-in and implement on a cross-platform operating system.

This section introduces the issue of interest and the significance of this research study. It provides details of the research problem and the research questions to be resolved together with the precise research objectives. It also summarises the existing literature and clarifies the main contributions of this research.

2. Online user decision support system protection against phishing attack using deep learning algorithm

This section contains a review of the literature on the topic under study, namely phishing detection schemes. It also discusses the focus of the research by critiquing the relevant existing research methods and summarising their findings as well as their strengths and weaknesses. It then discusses appropriate provision for the phishing detection problems and how to resolve them.

Big data has enabled machine learning algorithms to discover more fine-grained patterns and to make more accurate and timely predictions than ever before [10]. Deep learning techniques are used for object identification in images, the transcription of voice into text, matching news items and products with user interests and presenting relevant search results [11]. Deep learning architectures are composed of non-linear operations in multiple levels, such as neural networks (NNs) with hidden layers, or of complicated relational methods in reusable approaches [13]. The deep learning concept started with the study of artificial NNs [14], and it has become an active research area in recent years. In a standard neural network (NN), neurons are used to produce real-value activations, and with the adjustment of weights, the scheme behaves as required. Moreover, training the ANN with backpropagation makes it useful with gradient descent algorithms which have played a vital role in the model in the past decades. Although training accuracy is high with back-propagation, when it is applied to testing data, its performance might not be satisfactory [15].

Yi et al. (2018) designed two sets of features for web-phishing interaction features and original content. They also developed a scheme based on a deep belief network (DBN). The test, which included using real IP flows from an Internet service provider (ISP), indicated that the proposed DBN-based model was able to achieve an approximately 90% true positive rate. Also, in the area automotive proposed in [16] in which a deep NN was used to assist the driver in the aspect of traffic light classification, the techniques were used to develop a system to assist in driving. Currently, machine learning is continuously demonstrating its effectiveness in an extensive range of applications. The most common form of machine learning, whether deep or not, is supervised learning [12]. Also, Le et al. (2018) proposed a solution called URLNet, which is an end-to-end deep-learning framework for learning non-linear malicious URLs by detecting it from the URL. They applied a CNN to both the words and characters of the URL features to learn the URL embedding in a jointly optimised framework. This approach allowed their model to capture several types of semantic data, which would not have been possible using existing schemes. They also presented advanced word-embeddings to solve the problem of too many rare words being observed in a classification task [17]. They conducted their experiments on a large-scale dataset and demonstrated that their proposed method gave a strong performance that was better than that of an existing method. The approach has two branches; the first branch has a character-level CNN where character-level embedding is used to represent the URL. The second branch contains a word-level CNN where word-level embedding is used to represent the URL. Thus, word-embedding itself is a mixture of character-level embedding and individual word-embedding. Their approach works in such a manner that it does not require any expertise.

Below are some of the advantages of deep learning algorithms [15]:

Unsupervised Learning: It has robustness by getting most of its connecting structure in other to observe data, which is crucial in other to limit an enormous number of tasks and if the upcoming tasks are not known on time.

1. **Unlabelled Data**: It can learn from mostly from unlabelled data. This means that it can work in a semi-supervised situation, where not all the dataset has comprehensive and correct semantic tags.

2. **Develop Interactions**: It can exploit interaction that are existing across a vast number of tasks. These interactions exist because all that the algorithm task offer is a diverse view of the same underlying reality.

3. **Multifaceted Learning**: It can learn from complex with highly varying function with several disparities much higher than the number of training instances.

4. **Huge Dataset**: It can learn from a massive dataset of features and can compute the training data in a short period with several linear examples.

However, there are some challenges associated with deep learning algorithms regarding the issue of the data used [18], as follows:

1. **Unbalanced data**: This is an issue that occurs in learning and mostly happens during classification if there are more features of some class than others. This issue can be resolved by using some techniques that focus on the data level or the classifier level.

2. **Inadequate data for learning**: This is an issue that occurs when a limited amount of data is available for cross-validation methods which are mostly applied by dividing the available data into two sets, one for learning and the other for validation, in order to check the behaviour of the network. However, to gain a better knowledge of the network, the size and features may be modified for training and evaluating the various aspects of the network.

3. **Overflow of data**: This problem occurs in big data because the generation of data is growing exponentially, and it is forecast that the information contained big data will continue to increase daily.

4. **Partial data**: Sometimes, a collection of data is used for solving a particular task, but the data becomes partial when some of it is lost or because some of its variables or features are unidentified. To resolve this issue, it is necessary to approximate missing values and then discover the relationship between the identified and unidentified data. There some methods based on NNs [15] and some other approaches that can be used to solve the problem.

5. **High measurement:** Information in the real-world application is often overflowing from the determination of a specific problem point of view which can be handled by the algorithm.

Due to the growth in cyberspace technology, computer users have a significant role to play in making the Internet a safer place for everyone because cyber-attacks are targeted at achieving either financial or social gain [19] to the detriment of the user. On the other hand, some people undertake phishing activities for fun and a sense of accomplishment rather than for financial or social gain, but can also have adverse consequences for the user [1].

Phishing awareness has been improved through the development and use of online game training and email-based training to combat phishing attacks [20]. The use of legislation is a direct measure to reduce phishing by tracking and arresting those who are involved in this criminal activity. The US was the first nation to use laws to combat illegal cyber activities, and many cyber attackers have been arrested and arraigned. The main issue with this approach is the effectiveness of the laws as it is challenging to trace phishing attacks. Fraudulent websites naturally migrate quickly from one server to another. Also, an average phishing website is online for less than 48 hours [21]. Hence phishing attacks are committed very quickly and, subsequently, the criminals who commit these attacks also quickly disappear into cyberspace. The other issue is that many laws are applied only when the damage has been done, and the online user has already been defrauded as a result of phishing attacks. A great deal of background knowledge and experience of phishing and an enormous amount of related information was gained during this development. The use of high-quality datasets in phishing detection classification plays a significant role in building phishing model classifiers [22].

2.1 Long short-term memory (LSTM)

The LSTM algorithm Long short-term memory is based on the recurrent neural network (RNN), which is used to recognise the occurrence of patterns in time series and which also uses error flow in its analysis. However, the LSTM architecture was developed to overcome the shortfalls in RNN, which is a highly non-linear recurrent network with multiple gates and propagative feedback [23]. An LSTM layer contains a set of recurrently connected blocks, known as memory blocks. These blocks can be a look-alike version of memory chips in a digital system. Hence, each of the blocks includes one or more repeatedly connected memory cells and contains three multiplicative units, namely, the input, forget gate and the output, which provide non-stop analogues of the read, write and reset functions for the block cells [24]. The LSTM network has achieved excellent results in character recognition applications [23]. It has also been used extensively in the analysis of handwriting recognition, speech recognition and polyphonic music modelling, where the results have shown that its usage leads to an improvement in standard detection analysis with variance in the parameter [25]. It has also been used in language modelling to analyse speech in a speech recognition system, where it was found to show an improvement in confusion over the RNN [24].

2.2 Convolutional neural network (CNN)

In recent years, the convolutional neural network (CNN) has seen massive adoption in computer vision applications [26]. In the area of object recognition, CNN has also been used for feature extraction [27]. The CNN belongs to the family of multilayer NNs that are developed for use with two-dimensional data, such as videos and images [28]. CNN is one of the most prominent deep-learning methods where numerous layers are trained using a rigorous methodology.

As mentioned above, CNN has also been shown to be highly effective in computer vision applications [18] and is, therefore, commonly used for that purpose. The CNN contains an input layer, convolution layer, pooling layer, fully connected layer, and output layer. The input layer holds the raw image values; the convolutional layer computes the output of the node that is connected to local regions in the input layer; the pooling layer performs a down-sampling process along the three-dimensional dimensions; the fully connected layer calculates the session scores, and the output layer produces the results. Currently, three main techniques are used in CNN for image classification:

1. Unsupervised pre-training of the CNN with supervised fine-tuning,

2. Transfer learning by fine-tuning the CNN models that have been pre-trained on a natural image dataset and

3. Training the CNN from scratch using available pre-trained features [12].

2.3 Developing the IPDSS anti-phishing tool

This section presents the development of the online plugin model of the IPDSS. The development of the tool was performed based on traditional feature engineering, plus the classification algorithm methodology presented in previous section. Features were created based on the URLs, image features and website elements. The CNN and LSTM classifier were trained using one million URLs and over 10,000 images to build the model. A Toolbar concept was developed using a deep learning (DL) algorithm against legitimate, suspicious and phishing websites. The results showed that a voice-generating user warning interface with a green colour status and a text showing a warning was generated within 25 seconds before the page loaded to give the user a warning.

Due to the advances in technology and the adoption of new techniques, phishers have been able to improve their forged websites so that they now have high similarity with legitimate sites in terms of content. In tests, the current state-of-the-art solutions have been able to obtain 70–98% accuracy (see **Table 1**) in identifying

Status	No. of websites	Accuracy %	TP%	TN %	Average result %
Phishing websites	1000	93.5%	93.8%	6.2%	93.28%
Suspicious websites	100	94.5%	94.8%	5.2%	
Legitimate websites	1500	91.8%	92%	8%	

Table 1.
Test results for IPDSS by toolbar application.

legitimate website. However, these solutions must perform well in the real world, so there needs to be a significant improvement of 0.5% or higher [29]. Moreover, their level of accuracy in identifying suspicious websites should be higher still, and their accuracy in detecting phishing websites should be even higher [30].

The IPDSS scheme extractor algorithm is used to extract the necessary elements from the website's user is visiting. The extracted features were used to compare with knowledge model to determine whether the websites are phishing, suspicious or legitimate. The three modules user warning interface has:

i. A red colour status and voice generation with text directive warn the user if the requested site is a phishing web page,

ii. An amber colour status and voice generation with text directive warn the user if the requested site is a suspicious web page and

iii. A green colour status and voice generation with text directive show the user that the requested site is a legitimate web page.

2.4 Testing the IPDSS anti-phishing toolbar

To evaluate the toolbar concept, it was tested on 2600 websites including legitimate, suspicious and phishing websites. First, it was tested on 1000 phishing websites. The LSTM-CNN algorithm runs in the background as a knowledge module. When a URL is typed into the address bar (**Figure 1**), the algorithm inspects whether the requested website is a phishing link by comparing the current URL

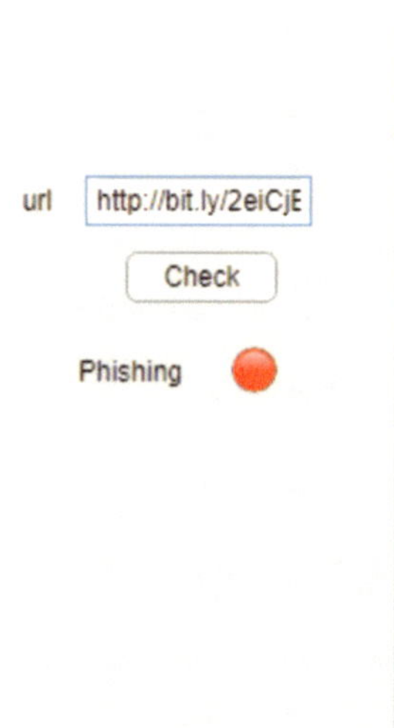

Figure 1.
Application interface for legitimate URL check.

against the stored features in the deep learning classification algorithm. If a match is detected, and it is a phishing site, in order to alert the user a red colour status with a voice-operated user warning interface is activated and a text is generated showing that the status of the URL is "phishing".

The above procedure was repeated up to 1000 times with different URLs, so all the phishing URLs were tested. The performance of the toolbar in each case was observed and recorded, and besides, screenshots were taken to validate the results. An example of a screenshot of a phishing website result is shown in **Figure 1**. This part of the experimental effort was carried out over 8 hours per day for five consecutive days. As regards the time-based assessment of the toolbar's ability to detect a phishing website, the voice-generating user warning interface with a red colour status and a text showing an alert were generated within 25 seconds to warn the user before the page loaded.

The toolbar also evaluated on 100 suspicious URLs. As previously mentioned, the LSTM-CNN algorithm runs in the background as a knowledge module. The same procedure is followed as in the testing of the toolbar on phishing websites that described in the previous section, but in this test, the algorithm checks whether the URL requested is a suspicious website by relating the newly typed URL against the stored features in the IPDDS. If a match is detected, and it looks like the URL is a suspicious website, the user warning interface included in the model shows an amber colour status and, besides, a text description is generated stating that the URL is "suspicious" (**Figure 2**) in order to alert the user to exercise caution. This process was repeated 500 times on all 100 URLs and the performance was observed and recorded (**Table 1**). An example of a screenshot of suspicious website results shown in **Figure 2**. This task required 8 hours per day over two days to perform because the finding shows that there is a little and a reasonable number of suspicious online websites which make this challenging task as they are short-lived. As regards the time-based assessment of the toolbar's performance in identifying a suspicious website, the voice-generating user warning interface with an amber colour status and a text showing a warning were generated within 25 seconds to alert the user before the page loaded.

The IPDSS was also tested on 1500 legitimate URLs. As stated above, the LSTM-CNN algorithm runs in the background as a knowledge module. The same procedure as that used to test the toolbar's performance on phishing and suspicious

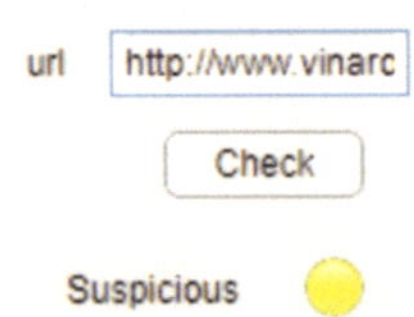

Figure 2.
Application interface for suspicious URL check.

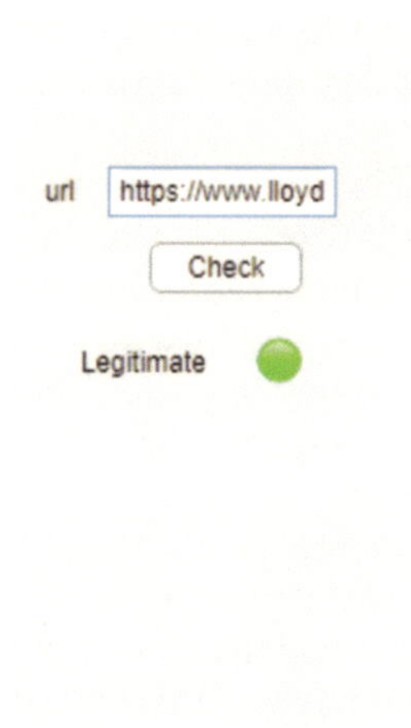

Figure 3.
Application interface for phishing URL check.

websites was used, but in this instance, the algorithm checks whether the URL that has been requested is a legitimate website by relating the newly typed URL in text box against the stored features in the IPDDS. If no match is found, then it is a legitimate website, and the user warning interface displays a green colour status (**Figure 3**). At this point, it is safe for the user to continue in their task with peace of mind that the site to which they are submitting their confidential information is legitimate.

In the experiment, this procedure was repeated 600 times with validation dataset consisting of URLs so that most the URLs were tested to validate the performance of the toolbar and in each case, the result was observed and recorded (**Table 1**). **Figure 3** shows an example of a screenshot of one of the results produced by the toolbar for a legitimate site. As regards the time-based assessment of the toolbar's ability to detect a legitimate website, the voice-generating user warning interface with a green colour status and a text showing the result was generated within 25 seconds before the page loaded.

Overall, the toolbar was able to achieve an average accuracy of 93.28%, as shown in **Table 1**. Then in **Table 1** column 4 roll 2, shows the performance of the phishing detection with 93.8% true positives and in column 5 roll 2, 6.2% true negative this has taken into consideration using 1000 phishing URLs with an accuracy of 93.5% in column 3 roll 2. Also, the toolbar achieved 94.5% accuracy shown on column 3 roll 3, with 94.8% true positives column 4 roll 3 and 5.2% true negative in column 5 roll 3 when tested on 100 suspicious datasets. Meanwhile, when the plugin is tested on 1500 legitimate websites, the phishing detection toolbar achieved 91.8% accuracy column 3 roll 4, was recorded with true positives of 92% column 4 roll 4 and 8% real negative in column 5 roll 4. However, accuracy varies from a minimum of 91% to a maximum of 94%, which caused significant variation in the accuracy results across the testing datasets.

3. Conclusion

This development also explored the efficacy of the deep learning approach, which is part of the set objective to explore relevant algorithm for the detection of phishing, this revealing the advantages and disadvantages of both the convolutional neural network (CNN) and long short-term memory (LSTM) methods. On the one hand, the LSTM+CNN algorithm was also used to develop an offline approach for phishing detection but had a smaller detection accuracy of 93.28% compared to that of the ANFIS algorithm.

The reduction in the number of features makes this much faster in terms of time-to-prediction. The protection aspect of the solution is implemented via a user warning interface with various colours representing the category of detection. A green colour indicates a legitimate site, whilst an amber colour represents suspicious ones, and a red colour indicates a phishing site. There is also an audible (voice) warning of relevance to a visually impaired person. The protection interface also advises the user on what to do next such as to terminate the process if it discovers that the site is phishing or suspicious.

The development reflects the effectiveness of the hybrid features approach using CNN, and the LSTM deep learning algorithm is an essential driver to the high model performance. This chapter has contributed to the anti-phishing detection research by present the use of a hybrid feature which include image, frame and text. These three sets of input have just been introduced as single hybrid features for the first time. The three elements are used because they represent the whole structure of a website. Although the scheme performed well, parameter tuning influenced the algorithm in a positive way, and it must be pre-specified to solve a given problem. Ultimately online user confidence will increase in performing transactions online.

The main conclusion of applying the IPDSS approach that is in this development achievement an excellent classification accuracy of 93.28% for identifying phishing websites.

Author details

Moruf Akin Adebowale
School of Computing and Information Science, Anglia Ruskin University, Chelmsford, UK

*Address all correspondence to: akin_mama@hotmail.com

References

[1] H. Sharma, E. Meenakshi, and S. K. Bhatia, "A comparative analysis and awareness survey of phishing detection tools," presented at the 2nd IEEE International Conference on Recent Trends in Electronics, Information & Communication Technology (RTEICT), Bangalore, India, 19-20 May 2017, 2017.

[2] N. A. G. Arachchilage, S. Love, and K. Beznosov, "Phishing threat avoidance behaviour: An empirical investigation," *Computers in Human Behavior,* vol. 60, no. 2016, pp. 185-197, 2016, doi: http://dx.doi.org/10.1016/j.chb.2016.02.065.

[3] S. Purkait, "Phishing counter measures and their effectiveness – literature review," *Information Management & Computer Security,* vol. 20, no. 5, pp. 382-420, 30 September 2018 2012, doi: doi:10.1108/09685221211286548.

[4] A. Upadhyaya, "Design & development of a plug-in for a browser against phishing attacks," *International Journal of Emerging Technology & Advanced Eng.,* vol. 2, no. 3, pp. 105-111, March, 2012 2012.

[5] Hu J et al. Detecting Phishing Websites Based on the Study of the Financial Industry Webserver Logs. presented at the 3rd International Conference on Information Science and Control Engineering (ICISCE), Beijing, China, 8-10 July. 2016;**2016**

[6] A. Y. Daeef, R. B. Ahmad, Y. Yacob, and N. Y. Phing, "Wide scope and fast websites phishing detection using URLs lexical features," in *3rd International Conference on Electronic Design (ICED)*, Phuket, Thailand, 11-12 Aug. 2016 2016: IEEE, pp. 410-415, doi: 10.1109/ICED.2016.7804679.

[7] Hong J. The state of phishing attacks. Communications of the ACM. 2012;**55**(1):74-81

[8] Tan CL, Chiew KL, Wong K, Sze SN. PhishWHO: Phishing webpage detection via identity keywords extraction and target domain name finder. Decision Support Systems. 2016;**88**:18-27. DOI: 10.1016/j.dss.2016.05.005

[9] O. K. Sahingoz, E. Buber, O. Demir, and B. Diri, "Machine learning based phishing detection from URLs," *Expert Systems with Applications,* vol. 117, no. 2019, pp. 345-357, 01 March 2019 2019.

[10] L. Zhou, S. Pan, J. Wang, and A. V. Vasilakos, "Machine learning on big data: Opportunities and challenges," *Neurocomputing,* vol. 237, no. 2017, pp. 350-361, 12 January 2017 2017.

[11] Tyagi I, Shad J, Sharma S, Gaur S, Kaur G. A Novel Machine Learning Approach to Detect Phishing Websites. presented at the 5th International Conference on Signal Processing and Integrated Networks (SPIN), Noida, India, 22-23 Feb. 2018;**2018**

[12] W. Yao, Y. Ding, and X. Li, "Deep Learning for Phishing Detection," in *Intl Conf on Parallel & Distributed Processing with Applications, Ubiquitous Computing & Communications, Big Data & Cloud Computing, Social Computing & Networking, Sustainable Computing & Communications (ISPA/IUCC/BDCloud/SocialCom/SustainCom)*, Melbourne, Australia, 11-13 Dec. 2018 2018: IEEE, pp. 645-650, doi: 10.1109/BDCloud.2018.00099.

[13] G. Montavon, W. Samek, and K.-R. Müller, "Methods for interpreting and understanding deep neural networks," *Digital Signal Processing,* vol. 73, no. 2018, pp. 1-15, 24 October 2017 2018.

[14] Vazhayil A, Vinayakumar R, Soman K. "Comparative Study of the Detection of Malicious URLs Using Shallow and Deep Networks," presented

at the 9th International Conference on Computing, Communication and Networking Technologies (ICCCNT), Bangalore. India. July 2018;**10-12**:2018

[15] W. Liu, Z. Wang, X. Liu, N. Zeng, Y. Liu, and F. E. Alsaadi, "A survey of deep neural network architectures and their applications," *Neurocomputing,* vol. 234, pp. 11-26, 2017/04/19/ 2017, doi: https://doi.org/10.1016/j.neucom.2016.12.038.

[16] CireşAn D, Meier U, Masci J, Schmidhuber J. Multi-column deep neural network for traffic sign classification. Neural Networks. 2012;**32**:333-338, 2012

[17] H. Le, Q. Pham, D. Sahoo, and S. C. Hoi, "URLnet: Learning a URL representation with deep learning for malicious URL detection," presented at the arXiv preprint arXiv:1802.03162, Washington, DC, US, 2 March 2018, 2018.

[18] Y. Guo, Y. Liu, A. Oerlemans, S. Lao, S. Wu, and M. S. Lew, "Deep learning for visual understanding: A review," *Neurocomputing,* vol. 187, no. 2016, pp. 27-48, 26 November 2015 2016.

[19] Arachchilage NAG, Love S. Security awareness of computer users: A phishing threat avoidance perspective. Computers in Human Behavior. 2014;**38**:304-312, 2014

[20] Arachchilage NAG, Love S. A game design framework for avoiding phishing attacks. Computers in Human Behavior. 2013;**29**(3):706-714. DOI: 10.1016/j.chb.2012.12.018

[21] A. Oest *et al.*, "Inside a phisher's mind: Understanding the anti-phishing ecosystem through phishing kit analysis," presented at the APWG Symposium on Electronic Crime Research (eCrime), San Diego, CA, USA, 15-17 May 2018, 2018.

[22] Zareapoor M, Seeja K. Feature Extraction or Feature Selection for Text Classification: A Case Study on Phishing Email Detection. International Journal of Information Engineering and Electronic Business. 2015;**7**(2):60-65. DOI: 10.5815/ijieeb.2015.02.08.

[23] T. M. Breuel, A. Ul-Hasan, M. A. Al-Azawi, and F. Shafait, "High-performance OCR for printed English and Fraktur using LSTM networks," in *12th International Conference on Document Analysis and Recognition*, Washington, DC, USA, 25-28 Aug. 2013 2013: IEEE, pp. 683-687.

[24] M. Sundermeyer, R. Schlüter, and H. Ney, "LSTM neural networks for language modeling," in *Thirteenth annual conference of the international speech communication association*, Portland, OR, USA, 9-13 September 2012 2012: ISCA, pp. 194-197.

[25] Greff K, Srivastava RK, Koutník J, Steunebrink BR, Schmidhuber J. LSTM: A search space odyssey. IEEE transactions on neural networks and learning systems. 2017;**28**(10): 2222-2232

[26] Y. Yu, Z. Gong, P. Zhong, and J. Shan, "Unsupervised Representation Learning with Deep Convolutional Neural Network for Remote Sensing Images," in *International Conference on Image and Graphics*, Cham, 2017: Springer, pp. 97-108.

[27] Z. Xu, S. Li, and W. Deng, "Learning temporal features using LSTM-CNN architecture for face anti-spoofing," in 2015 *3rd IAPR Asian Conference on Pattern Recognition (ACPR)*, Kuala Lumpur, Malaysia, 3-6 November 2015 2015: IEEE, pp. 141-145.

[28] Arel I, Rose DC, Karnowski TP. Deep Machine Learning - A New Frontier in Artificial Intelligence Research [Research Frontier]. IEEE Computational Intelligence Magazine. 2010;**5**(4):13-18. DOI: 10.1109/MCI.2010.938364

[29] Shirsat SD. "Demonstrating Different Phishing Attacks Using Fuzzy Logic," presented at the Second International Conference on Inventive Communication and Computational Technologies (ICICCT), Coimbatore. India. April 2018;**20-21**:2018

[30] (2018). *Phishing attacks: defending your organisation*. [Online] Available: https://www.ncsc.gov.uk/phishing